Physiology and Physical Activity

Harper's Series on Scientific Perspectives of Physical Education, edited by Rainer Martens

PHYSIOLOGY AND PHYSICAL ACTIVITY

Brian J. Sharkey

Harper & Row, Publishers
New York Evanston San Francisco London

Sponsoring Editor: Joe Ingram
Project Editor: Ralph Cato
Designer: Michel Craig
Production Supervisor: Stefania J. Taflinska

Physiology and Physical Activity
Harper's Series on Scientific Perspectives of Physical Education

Library of Congress Cataloging in Publication Data

Sharkey, Brian J. 1937–
Physiology and physical activity

(Harper's series on scientific perspectives of physical education)
1. Human mechanics. 2. Exercise. I. Title.
[DNLM: 1. Exertion. 2. Movement. 3. Physical education and training. WE103 S531p]
QP301.S46 612'.76 74-5590
ISBN 0-06-045965-4

CONTENTS

SERIES PREFACE

As has occurred in several other fields, the knowledge explosion in physical education and its collateral sciences—the biological and behavioral sciences—warrants a different approach for conveying introductory material. Traditionally, a basic course has used a single text that superficially covers everything, but sacrifices depth. The alternative is a text that covers one area in depth, but ignores many essential topics. Neither approach is adequate. Physical education has become too diverse for any one individual to write about all areas with complete authority. The developments in the behavioral and biological areas of the field have been too rapid.

The Scientific Perspectives of Physical Education series essays to remedy the instructor's dilemma of choosing between breadth and depth. It is composed of five volumes, representing the fields of physiology, kinesiology, psychology, social psychology, and sociology. In that order, the series moves from a biological emphasis to a social emphasis. Each science is included because of its obvious relationship to the other fields, as well as for its growing body of knowledge, which is a function of the amount of scholarly activity occurring within it. And, of course, each field is included because of its relevance for the diversified physical education profession.

The Scientific Perspectives of Physical Education series, then, is a direct outgrowth of the scholarly activity occurring in physical education and collateral sciences. Physical educators are now realizing the importance of reaching the introductory, as well as upper level, students with high-quality, well-written, interesting material. Rather than material based on speculation and conjecture, information that highlights the continuing and exciting search for new knowledge is being sought. This series is our attempt to place such material in the hands of physical educators.

Each volume is written by an individual actively engaged in research in this field. The approach of the series has been to focus on the reciprocal relationship between physical activity or sport and the subject area. More explicitly, the influence of physiological, biomechanical, psychological, social psychological, and sociological variables on physical activity or sport are considered, as well as the influence of engaging in physical activity or sport on these same variables.

Although the material is written for the introductory student in physical education, it presumes that the student has a basic understanding of the various disciplines represented. Within each area we hope to convey what is known today, what is currently being studied, and what needs to be understood. Emphasis is placed on content rather than method. Although method dictates the quality of the content, the authors assume responsibility for discriminating between quality and lesser research. Documentation of material has been used sparingly, primarily for recent work, in order to retain readability.

By no means is the series to be considered definitive within each area or for the entire field. Much of our knowledge is incomplete; our understandings are limited. But it is necessary to remember that the acquisition of knowledge in each area is an ongoing process. Each author has attempted to tell it as it is, not as what he hopes or conjectures it to be.

The series offers some flexibility because each volume is self-contained and may be used separately, or as a partial or complete set for a foundations course. Although written for the introductory undergraduate student, dependent upon the student's background and the curriculum, the series may be useful at more advanced levels.

Rainer Martens

PREFACE

You should not be surprised to learn that physiology and physical activity are closely related. Physiological factors influence and limit our activity, and physical activity may influence certain physiological factors. Exercise has acute or immediate effects as well as chronic or training effects. The study of these relationships—exercise physiology—provides the well-documented evidence that forms the foundation of physical education and sport programs.

This monograph deals with the physiology of human movement, and the applications of that study to the world of physical education and sport. I have attempted to distill the essence of an immense volume of literature and present it in a straightforward fashion to those with a basic but limited background in chemistry and physiology. To do this, I have drawn on my experience as a teacher, advisor, and researcher. I have consulted with authorities and colleagues, attended scientific sessions, and scoured journals and other available reference materials. But most of all, I have allowed my experiences with students, teachers, coaches, children, and adults to direct and focus this monograph toward relevant questions. References to basic physiology are limited in an attempt to retain that focus. Such references are included, when necessary, to provide the background needed for the understanding of problems in exercise physiology. Thus we will lightly sketch the outlines of the discipline (exercise physiology) as we focus on the problems you will face as a professional.

Our text is divided into three parts, each containing four chapters. Each chapter is short enough to be digested in one sitting, and each contains sufficient documentation to stand by itself as a supplementary source. The focus of each part is presented in a short introductory statement. Tables and graphs provide additional insight, and the Appendixes include some useful information and activities for the teacher, coach, or fitness enthusiast.

It was difficult to squeeze this immense and complicated subject into monograph form. The task allowed little time for nostalgic journeys into the memorable history of the field. I had to suppress the urge to fully detail and document the development of a concept. Because I had to make decisions regarding relevance, you should know something of my position or bias. As a researcher I am committed to a search for the facts, rather than a justification of current practices. As a professional physical educator I am committed to optimal development for all, rather than maximal development of a few. As a researcher and a professional I have considered it my responsibility to provide service in the form of possible solutions to the problems encountered in the real world. Therefore, I hope that this monograph enables you to better appreciate the fascinating discipline of exercise physiology and to understand its relevance to your professional goals. I can only hope that you will discover the beauty inherent in the subject and continue to seek its delights and frustrations throughout your career as I have done.

Before you begin, allow me one final comment in an attempt to ease your journey through this monograph. You will be confronted with numerous theories and may wonder why researchers are unable to provide positive answers to your questions. Remember that the typical exercise physiologist goes to work at the university, teaches several classes, advises students, serves on faculty committees, keeps up with the research literature, conducts research, and writes research reports. We also attempt to get some exercise every day. To be sure, a limited number of researchers work in labs where the entire day is devoted to the advancement of the science. Whatever the situation, each researcher works to develop hypotheses and test theories in a limited portion of the discipline. Each is limited by time, money, and instrumentation. As new techniques and instruments become available, new methods are found to answer the problems that have been blocking the advancement of knowledge in the field. On occasion, breakthroughs take place and knowledge takes a giant step. More typically, however, the steps are short and often painful. Sometimes we take one step backward in order to take two steps forward.

If you fail to find all the answers to your questions, do not lose hope. The information you seek may be found in the next edition of *Medicine and Science in Sports* or the *Journal of Applied Physiology*; it may be reported at the next scientific meeting, or you may even read about it in *Sports Illustrated* or *Readers Digest*. Any lively science is undergoing a period of dramatic change, and it is just this uncertainty that makes physiology and physical activity so exciting.

Brian J. Sharkey

Physiology and Physical Activity

PART I PHYSICAL ACTIVITY: INITIATION OF MOVEMENT

We will begin our consideration of the relationships between physiology and physical activity with a brief review of the conscious initiation and control of muscular contractions. We will then consider the nature of the contractile process, the fuels for contraction, and the possible types of contractions. If all that sounds a bit heavy, keep in mind that we shall cover that territory in order to deal with questions concerning performance and training, strength and power, and endurance and fatigue. Thus, we will be dealing with the substance of the discipline in order to better face the problems encountered within the profession.

We shall see that movement is initiated on command of the motor cortex and controlled via an elaborate and elegant system of feedback. We shall observe that the muscular contraction can be powerful and brief or less forceful but sustained, and that the nature and intensity of the contraction dictates the fuel used for energy. We shall also attempt to provide some practical guides for the training of muscular strength and endurance.

But keep in mind that muscular contractions, movement, and physical activity in general only result when you make the *decision to move*. The pleasures and possible values of physical activity come only to those who decide to move, and to engage in enjoyable and meaningful physical exercise. All the physiological facts you can learn will not budge the lethargic adult who has learned that exercise is painful and awkward, degrading and humiliating. On the other hand, many of us will continue to risk the dangers of skiing, mountain climbing, or other thrilling movement experiences we have learned to enjoy. While physical activity may produce profound physiological changes, it is likely that the immediate joy to be found in movement is a more potent force leading to the decision to move.

CHAPTER 1 THE CONTROL OF MOVEMENT

Before we begin our discussion of the conscious control of movement, let us consider sleep and the transition to consciousness. While asleep we seem to be relatively inactive; of course, the heart and breathing muscles continue to contract, and other involuntary muscles promote peristalsis, maintain sphincter control, and regulate blood flow throughout the body. Sleep seems to be characterized by alternating stages. One stage involves rapid eye movements (REM), and changes in heart rate, blood pressure, and muscle tone. This stage may serve as a rest period for the inhibitory nerve cells of the brain. It is usually accompanied by dreams, and if it is interrupted we become anxious and irritable. This REM sleep comprises about 20 percent of the night's total, while a deeper, quieter stage (non-REM or nREM) seems to provide the rest so necessary for our recovery from fatigue. Moderate physical activity seems to enhance the ability to fall into nREM sleep, without altering the time spent in REM sleep. Too little or too much exercise appears to result in sleep disturbance. (Hobson, 1968.) The amount of exercise required to enhance sleep seems to be dependent upon physical fitness as well as on one's previous exercise experiences.

Our passage from sleep to conscious awareness is mediated by the reticular activating system (RAS) located in the brainstem. The RAS, so intimately related to the activity of the cerebral cortex, is probably responsible for sleeping and wakeful states, as well as the focusing of attention to various portions of the cortex and the narrowing of attention (as in a game) to exclude unwanted sensory information (like the noise of the crowd). Thus, we move to a wakeful state and begin to focus on the events and demands of the day. From this point on, much of our movement will come under the direct conscious control of the higher centers of the brain. Why do we select one course of action and not another? What initiates that burst of nervous activity that puts us in motion?

HIGHER CENTERS AND THE CONTROL OF MOVEMENT

The decision to move may arise in the integrative area of the cerebral cortex (Fig. 1.1). It may receive thoughtful consideration in the prefrontal area (judgment), where it can be inhibited. A course of action is then selected and the action initiated by the motor cortex. Incoming sensory information could also initiate movement. A fist that is thrown in your direction, for example, would be detected visually and the information quickly relayed via the visual cortex to the integrative area. The motor area could quickly initiate the movement needed to block the incoming fist while subsequent action was being considered by other areas of the brain. Your previous experiences in fights, the size of your opponent, and the emotional and social setting could all influence your next movement decision, that is, whether to fight or to flee.

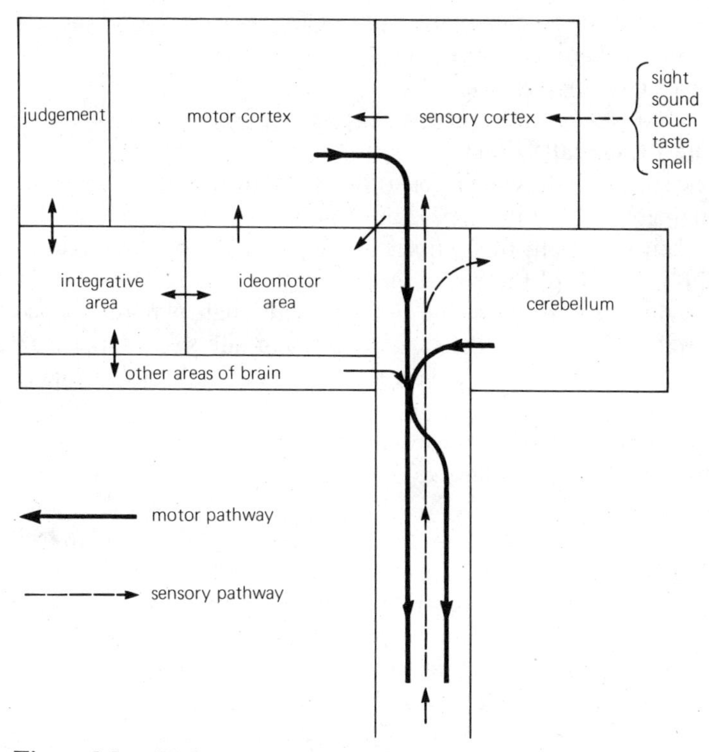

Figure 1.1. Higher centers and the control of movement. The decision to move may be initiated by an incoming stimulus or it may arise spontaneously within the brain. Upon consideration, a movement pattern is selected and then carried out with the constant feedback control arising from sensory receptors located in the muscles. Adjustments are made during the movement to make it coincide with our memory of the skill or to assist in the achievement of a desired goal (e.g., to strike a target area).

Movement is often preceded by a *decision*. In spite of what some have called a basic need to move, it appears that the need for physical activity varies among us all. Some need (or have learned to need) a great deal of physical activity; others seem to lead happy, healthy lives engaging in a minimal amount of activity. (Williams, 1967.)

Numerous structures are involved in the initiation and control of the impulses that ultimately converge on the final common path—the alpha (larger) motor neurons which directly control the activity of muscles. Let us briefly survey those mechanisms, beginning with those subject to our conscious control (see Fig. 1.1).

The Pyramidal System

The nerve cells which form the pyramidal system seem to originate in the motor areas of the cerebral cortex. However, it is likely that many of the fibers of the pyramidal system originate in other areas of the cortex, including some from sensory areas. Whatever their origin, the fibers pass through pyramid-shaped bundles in the medulla, cross to the opposite side of the cord in the lower medulla (80 percent of them cross), and terminate down the cord in connection with motor or connecting neurons. Thus, an injury or stroke on the left side of the cortex will be characterized by failure to initiate or control movement on the right side of the body.

The pyramidal system seems to initiate skilled, voluntary, discrete movements. However, to illustrate the complexity of the system and the danger of oversimplification, it should be mentioned that skilled movements may be possible when the system is damaged. But, in the intact human, it is the point of origin for the multitude of skilled voluntary movements which are manifest in the diverse forms of physical activity and sport.

The Extrapyramidal System

This system includes all other higher centers that exert an influence on the motor neurons. Included are extrapyramidal cortical fibers that evoke or inhibit complex movements or reflexes, the cerebellum and associated feedback mechanisms, and other brain structures that seem to be involved in the control of movement. Some authors also include the vestibular system because of its influence on posture and related reflex mechanisms. The extrapyramidal system seems to be concerned with the modification of movements, ranging from postural adjustments to subtle controls over the force and extent of contractions. This system only reacts to movement rather than initiating it. Further, it seems that the extrapyramidal system is concerned with the postural and other background adjustments needed to complete a voluntary movement as well as with facilitory and inhibitory influences governing the movement. It is one thing to punt a football (pyramidal system), and yet another to avoid falling on the turf after the kick (extrapyramidal system).

Spinal Reflex Mechanisms

Spinal reflex mechanisms may interact with the extrapyramidal system to influence, facilitate, or inhibit movements initiated by the consciously controlled pyramidal system. They can also initiate a variety of involuntary movements seemingly designed to provide rapid feedback on the state of contraction or to protect us from harm. While a detailed description of the complex mechanisms involved is outside the scope of this text, it is important to mention some fundamental aspects of reflex motor control.

Spinal reflex mechanisms depend upon *sensory information* such as muscle tension, the extent and direction of changes in tension, deformation of the skin, pain, heat, cold, and so on. The sensory information is relayed to the spinal cord via the dorsal root where it synapses with the motor neuron, thereby initiating a contraction (Fig. 1.2). Conversely, a smaller motor neuron in the cord (gamma motor neuron) is capable of exerting control over some sensory receptors in the muscles. Thus, motor output determines sensory input and sensory input determines motor output. (Thompson, 1967.) This system of dual control based on feedback allows us to adjust our contractions to fit the demands of the changing situation. Skilled movements are initiated by the motor cortex, modified by the extrapyramidal system, and finely tuned by the self-adjusting feedback (servomechanism) from spinal reflex mechanisms. However, Roberts, Smith, and Roberts (1970) have found that these spinal servomechanisms do not furnish feedback control during very rapid voluntary movements. During ballistic movements such as throwing, the golf swing, and the tennis serve the final common pathway of the movement may not be subject to alteration or adjustment. Thus, extremely talented performers probably achieve their effect more by the quality of the final common pathway than by their ability to adjust the movement in progress.

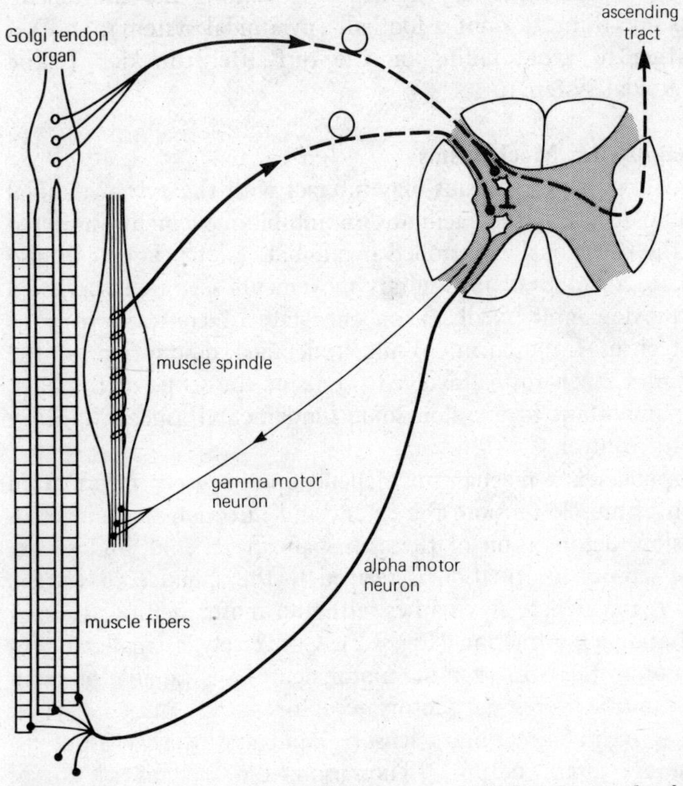

Figure 1.2. Sensory control of movement. Moderate stretch of the muscle stimulates the muscle spindle and acts to provide assistance via the alpha motor neuron (stretch reflex). Excessive stretch activates the Golgi tendon organ which serves to inhibit the contraction via the action of an inhibitory interneuron in the cord. Sensory information regarding the contraction reaches the brain by way of ascending tracts in the white matter of the spinal cord.

Even the so-called simple reflex is not so simple. It typically involves an interneuron, communication with an ascending tract, and possible inhibitory mechanisms (see Fig. 1.2). Reflexes may involve sensory information at one level of the cord and motor neurons at another (intersegmental reflexes). When higher brain centers are required to facilitate or to coordinate the spinal mechanisms, we speak of suprasegmental reflexes. Finally, reflexes often inhibit the action of antagonistic muscle groups (reciprocal innervation) in order to allow the completion of a movement.

We have quickly surveyed some of the principal structures and features involved in the control of movement. We have seen that voluntary and skilled movements begin in the motor cortex. Important facilitory or inhibitory reflexes, as well as postural support, come via the extrapyramidal system. Then the movement itself is adjusted in force and amplitude via facilitation or by inhibition provided from the sensory feedback of the spinal reflex mechanisms. Researchers have also demonstrated the influence of higher centers on the gamma motor neurons and resulting changes in muscle tension and sensory feedback. These topics are outside the scope of this survey.

SENSORY INFORMATION FROM MUSCLES

The sense receptors located in and about our muscles comprise the largest sense organ system in the body. Some forms of physical activity provide pleasurable sensations that lead to repetition of the activity. It seems likely that we can cultivate this somatic sense just as we develop the ability to appreciate sounds (e.g., music) or sights. Some of this muscle sense or kinesthesia is relayed to the sensory cortex while some seems to terminate in the cerebellum. Whatever its point of termination, it seems obvious that either the movement itself or the result of the movement provides pleasure, satisfaction, and meaning to the experience.

Let us consider the kinesthetic receptors and their mode of action. Your brain is able to decode the many incoming stimuli by the termination point of nerve fibers and by the frequency of impulses carried by the nerves. *Muscle spindles* indicate the degree of stretch on the muscle by the frequency of fiber discharge. These spindles consist of a connective tissue sheath that encloses some muscle fibers (intrafusal) and two sorts of sensory end-organs. Since they are arranged parallel to the skeletal muscle they serve, they are stretched when the muscle is stretched (Fig. 1.2). When stretched, the primary or annulospiral ending sends rapid messages to the motor neurons of the stretched muscle, thus initiating the stretch reflex. The secondary or flower-spray ending is also stretch sensitive, but it seems to offer information regarding length alone, while the primary ending signals both the length of the muscle and the velocity at which it is being stretched. As the muscle contracts, the discharge from the spindle receptors is diminished. At this point, the smaller gamma motor neurons will activate the intrafusal fibers and readjust the length of the spindle so it will be ready again to provide sensory information.

The *Golgi tendon organ* is activated by forceful stretch of the muscle, but its effect is inhibitory while that of the muscle spindle is facilitory. Thus, the resistance of the barbell during a contraction may tend to stretch the muscle and call forth assistance (facilitation) via the muscle spindle. However, should the weight prove quite heavy and cause a very strong stretch, the Golgi organ will try to bring about relaxation (inhibition) of the potentially dangerous contraction. The muscle spindle and Golgi tendon organ seem to have two functions: Both serve as kinesthetic receptors and send information concerning muscle tension to the central nervous system (CNS). However, while the spindle serves to facilitate contractions the tendon organ acts to inhibit those contractions that could result in injury.

When a muscle contracts, a further feedback mechanism is activated to keep the contraction from becoming *too* strong. The stimulus activating the muscle contraction also activates a Renshaw cell which serves to inhibit other motor neurons and the muscle fibers they excite.

The *pacinian corpuscle* is an onion-shaped receptor located deep in the tissue surrounding the joints. It is deformed by active or passive movements, and is an important indicator of limb position. These receptors—along with information supplied by free nerve endings, touch, heat and cold receptors in the skin, information supplied by sight and sound organs, and the vestibular information concerning acceleration and gravity—combine to inform us of our position in space, as well as the force, extent, and *quality* of our movement.

THE NEURON AND SYNAPSE

In our discussion of motor control and feedback we have assumed some understanding of the role of the neuron and communication in the nervous system. Let us briefly review the action of the neuron and synapse before we discuss their role in the learning of motor skills.

The single nerve cell or neuron is believed to be the basic functional unit of the nervous system. However, glial cells comprise 90 percent of the cells in the brain. Researchers are actively investigating the role of these cells that were once thought to be mere connective tissue. The billions of neurons in the nervous system are characterized by a unique ability to transmit impulses from place to place. This transmission comes about when a stimulus (usually a chemical one from adjacent neurons) changes the permeability of the cell membrane in the region of the cell body. The alteration in permeability allows the inflow of sodium ions, previously excluded from the neuron by an active mechanism called a *sodium pump*. As the ions flow inward, they alter the permeability of the adjacent area and the impulse travels along the length of the neuron (Fig. 1.3). When it reaches the axonal or terminal end of the neuron it activates the release of a chemical transmitter substance that serves to carry the message along to the next neuron, or to the tissue (e.g., muscle) that is the recipient of the message. Following the transmission of the impulse, the neuron pumps the sodium out and is again ready to carry another message.

Since each nerve cell body may receive many hundreds of chemically transmitted messages, and since each neuron may branch from a few to over a thousand axonal endings (e.g., the motor neurons may activate from 5 or 6 to 2000 individual muscle fibers), the number and variety of possible interconnections is staggering. A further complication exists since the chemical messages may either excite or inhibit the tissues served by the neuron. Excitation may be caused by one chemical and inhibition by another. While several substances have been suggested as possible transmitter substances in the CNS, it has been established that acetylcholine is the transmitter that activates the contractions of skeletal muscles. (Eccles, 1964.)

These chemicals may cause such an effect by decreasing the

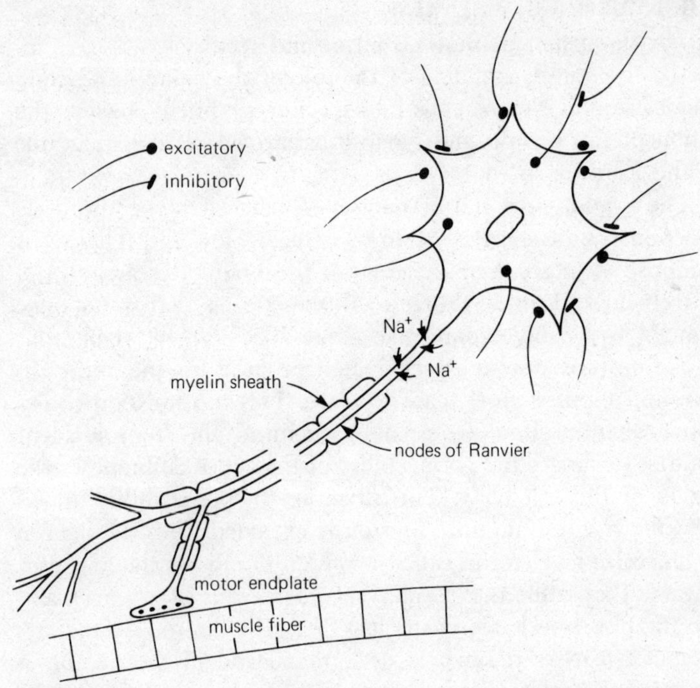

Figure 1.3. The motor neuron and the motor endplate. Excitatory stimuli result in an inward flow of sodium ions and the conduction of a stimulus to the muscle fibers served by the motor neuron. Inhibitory stimuli reverse the flow of sodium ions and block conduction of the impulse.

electrical difference between the inside and outside of the neuron (excitation) or by increasing the electrical negativity (inhibition). A nerve impulse is propagated when a sufficient amount of the excitatory chemical has altered the electrical state from —70 mv to the trigger point, about —60 mv. Since each nerve cell body receives numerous excitatory and inhibitory messages, the final decision to conduct or not to conduct is the result of the predominate message (Fig. 1.4). If each excitatory unit is balanced by an inhibitory unit, the electrical state is unchanged and no message is transmitted. If sufficient units of the excitatory substance are present, the electrical state is changed, so that the permeability is altered and the nerve impulse propagated.

It is worth noting that muscles deprived of nervous stimulation by denervation undergo different structural and metabolic changes than do muscles undergoing long periods of inactivity (e.g., a leg in a cast). The muscle with the intact nerve supply is much less affected by lack of use than is the denervated muscle. The so-called trophic function of the nerve seems to retard the process of atrophy, possibly by virtue of a subthreshold release of acetylcholine (AcCh) or some other chemical communication at the myoneural junction. (Guth, 1968.)

Let us now consider the metabolic demands of neurons and nerve tissue (discussed in greater detail in Chapter 4). Nervous tissue seems to be entirely dependent upon oxygen and blood

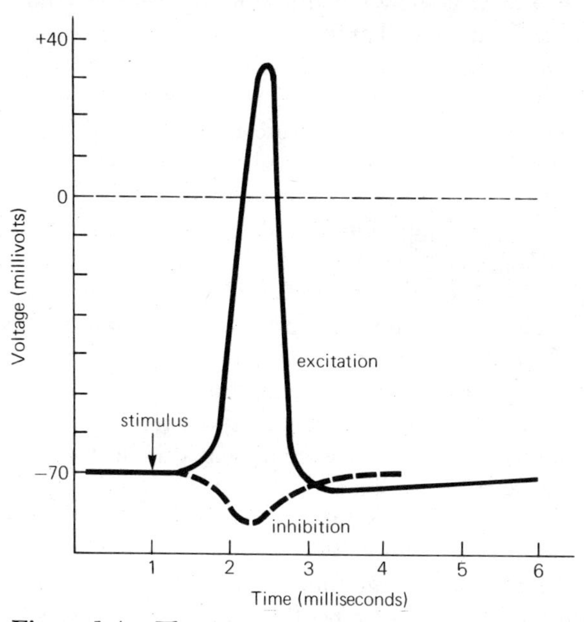

Figure 1.4. The action potential. Excitation of the neuron moves the electrical state toward the trigger point (−60 mv). Inhibitory stimuli increase the negativity of the membrane potential beyond the resting level (−70 mv).

sugar (glucose) for its energy supply. The brain is the first organ to react to oxygen or glucose deficiency, and it is possible that many of the subtle symptoms of fatigue are manifestations of these needs. This position differs with the traditional view which holds that the neuron is relatively indefatigable. For the moment, however, we should note that both oxygen and glucose become less available to the tissues during *prolonged* strenuous exertion.

MOTOR LEARNING

You need not consult a reference book to realize that we learn a variety of habits and motor skills through the process of repetition. Early attempts may be under our conscious control, but eventually the skill can be carried out rather automatically, almost as a reflexive action. The essential problem for the physiologist is to describe how repetition of a movement establishes a preferred pathway that ensures accurate future performance of the skill. Also, we would like to be able to describe the optimal conditions for the creation of the pathway to enable teachers and coaches to structure more effective learning situations. How are complex skills such as swimming, skiing, or bicycling stored in memory for future use? Why are we able to remember them when important cognitive skills (e.g., algebra) are often forgotten?

While the entire learning-memory problem is far from resolved, recent human and animal research does offer some clue to the physiological mechanism for skill learning. Deutsch (1971) has presented evidence supporting the late nineteenth-century concept that learning and memory result from a change in synaptic conductance. The evidence presented indicates that cholinergic synapses are modified, and that the postsynaptic membrane becomes increasingly more sensitive to AcCh during the learning process. Sometime after learning, the sensitivity begins to decline, leading to forgetfulness. Increasing the amount of learning (i.e., by practice or repetition) leads to an increase in conductance but does not increase the number of synapses involved.

Thus the repetition of a skill could lead to a decrease in the synaptic resistance and an increase in the probability that a certain final common pathway would be utilized. The frequent practice of motor skills may account for their excellent retention in contrast to many cognitive skills. The findings cited do not begin to explain all the problems of long- or short-term memory. These and associated problems will probably be explained, not in narrow physiological or psychological terms, but in concepts that combine recognition of the biophysical, biochemical, and the behavioral aspects of learning.

While the study of the structural and chemical aspects of learning is in its infancy, various other mechanisms have received considerable attention. They include the swelling of presynaptic terminal knobs, growth of new presynaptic knobs, and increased ribonucleic acid (RNA) synthesis. The integration of such concepts with the various behavioral manifestations of learning could assist the processes of hypothesis testing, theory building, and model construction. Whatever the approach, it is obvious that this is an essential area of research for those interested in the domain of physical activity. For clearly, the continued pursuit of physical activity is predicated on the success and pleasure derived therein. The development of effective methods for skill instruction can only be aided by a clearer understanding of how skills are learned, stored in memory, and made available for future use—when you make a decision to move and consciously select a preferred activity for the pleasures and benefits you have found it provides.

CHAPTER 2 THE CONTRACTILE MECHANISM

In Chapter 1 we omitted a discussion of the transmission of the nerve impulse to the skeletal muscle. Let us now follow the stimulus through the body of the muscle to the muscle fiber, where the actual shortening takes place. The motor neuron leaves the spinal cord via the ventral root and wends its way to the muscle. Muscles, such as the biceps brachii, are made up of bundles of muscle fibers that are separately wrapped in fibrous connective tissue. Each bundle, made up of thousands of individual fibers, is wrapped in a connective tissue sheath, as is each fiber within the bundle. Hence, each muscle fiber is effectively insulated from its neighbor. The well-insulated motor neuron enters the muscle and finds its way through the bundles until it reaches the region it is to serve. Here it branches from a few to over a thousand times, depending on the degree of motor control necessary. Motor neurons serving the delicate eye muscles activate only a few small muscle fibers, while those serving the powerful muscles of the leg branch profusely.

The muscle fibers served by each motor neuron comprise a *motor unit*, in such a way that a stimulus descending the nerve will activate all the fibers served. Just as each muscle fiber responds in an all or none fashion, each motor unit responds as best it can under the circumstances. Factors such as fatigue, oxygen or fuel supply, temperature, muscle length, and others can influence the production of tension. Recent investigations have provided some interesting insights into the structural and functional relationships of mammalian motor units.

In one such experiment, single motor neurons were stimulated to activate but one motor unit of the cat gastrocnemius muscle. Continued stimulation led to fatigue, and during the experiments various measures of force and fatigue susceptibility were recorded. Following stimulation, the muscles were quickly removed and prepared for histochemical analysis (i.e., the preparation of slides using various staining procedures). The analysis revealed three distinct types of motor units. They were labeled fast contracting with fast fatigue (white), fast contracting but fatigue resistant (red), and slow contracting. All of the muscle fibers in a given motor unit have the same histochemical profile and contractile properties. The white fibers are rich in glycogen but seemingly low in capillary supply, indicating dependence upon anaerobic (nonoxidative) energy sources. The red fibers seem to have both aerobic and anaerobic capabilities. They have a high-oxidative enzyme activity and a rich capillary supply. They also possess an abundance of glycogen. (Burke, Levine, and Zajac, 1971.)

In contrast to the animal studies, muscle biopsy samples taken from humans contain only *two* fiber types (Fig. 2.1):

Fast twitch, also called white, fast-contracting-fast fatigue, fast twitch glycolytic

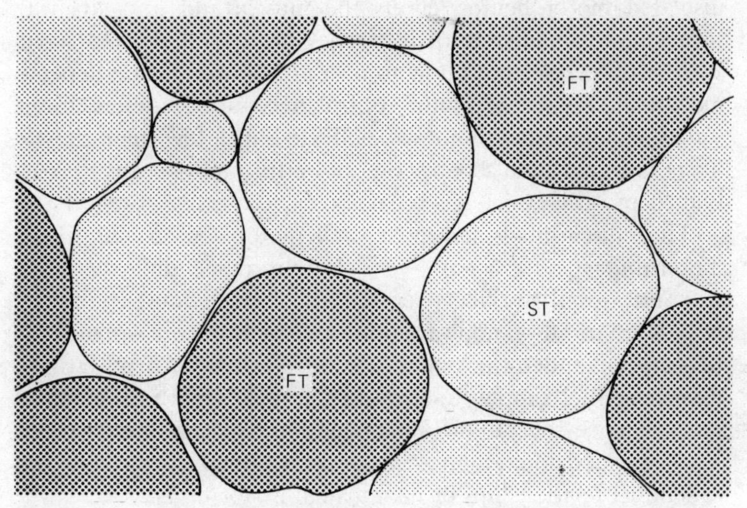

Figure 2.1. Fast- and slow-twitch fibers intermingle in human vastus lateralis biopsy samples. (From Gollnick et al., 1972.)

Slow twitch, also called red, fast-contracting-fatigue resistant, fast twitch oxidative-glycolytic

The red or slow-twitch fibers seem to predominate in the muscles of endurance athletes. (Gollnick et al., 1972.) One might expect weight lifters to have a preponderance of white fibers in their lifting muscles, however, a recent report indicates a composition approximating 50 percent of each (FT and ST). (Saltin, 1973.)

During short intense sprints the glycogen from the white fibers is depleted, indicating the involvement of these fast-twitch fibers in high-speed or power activities. During endurance runs the red fibers are the first to deplete. The white fibers are recruited only after the fatigue-resistant red fibers have severely depleted their energy stores. (Armstrong, Saltin, and Gollnick, 1973.)

It is likely that the distribution of white and red fibers in a muscle is under strong genetic influence. Thus, muscular endurance may be largely an inherited trait. However, experiments have indicated that the exchange of nerves from slow- and fast-twitch muscles reverses the function of the muscle fibers as well as their capillary profile. (Romanul, 1971.) This would indicate that the characteristics of the motor units are connected with the ways in which the units are used by the central nervous system (CNS) in movement. It also suggests that some potential for adaptability due to training resides within the muscle fibers. We will say much more concerning this point in Chapter 3.

CONTRACTION

The junction of the motor nerve and the muscle fiber is called the *motor endplate* or the *myoneural junction* (muscle-nerve junction). As an impulse reaches the end of the nerve, it activates the release of a quantity of the chemical transmitter substance acetylcholine (AcCh) into the microscopic space separating the endplate and muscle fiber. Receptor sites on the fiber react to the AcCh, and the membrane undergoes a change in permeability and depolarization (as found in the neuron). The AcCh is quickly removed by an enzyme called *cholinesterase,* but not before it has propagated a muscle action potential that spreads quickly along the length of the fiber. Just how the depolarization of the fiber leads to the initiation of contraction is not completely understood, but before we attempt to trace these events we must first consider the finer structure of the individual muscle fiber.

Fibers are made up of numerous myofibrils which contain the contractile proteins or filaments. Cell nuclei and mitochondria lie beside the fibrils which are enveloped in a tubular network (the sarcotubular system). The sarcotubular system assists in the rapid transmission of the action potential from the membrane surface to all the fibrils within the fiber. The organization of the filaments into sarcomeres is consistent for adjacent fibrils, giving the appearance of light and dark bands. The protein filaments—the thick myosin and the thinner actin—are arranged as is seen in Fig. 2.2. The area between two Z lines, called a sarcomere, is the functional component of the contractile mechanism. Thus, the message brought by the motor nerve, transmitted by AcCh, and somehow carried into the fiber via the sacrotubular system must somehow cause the

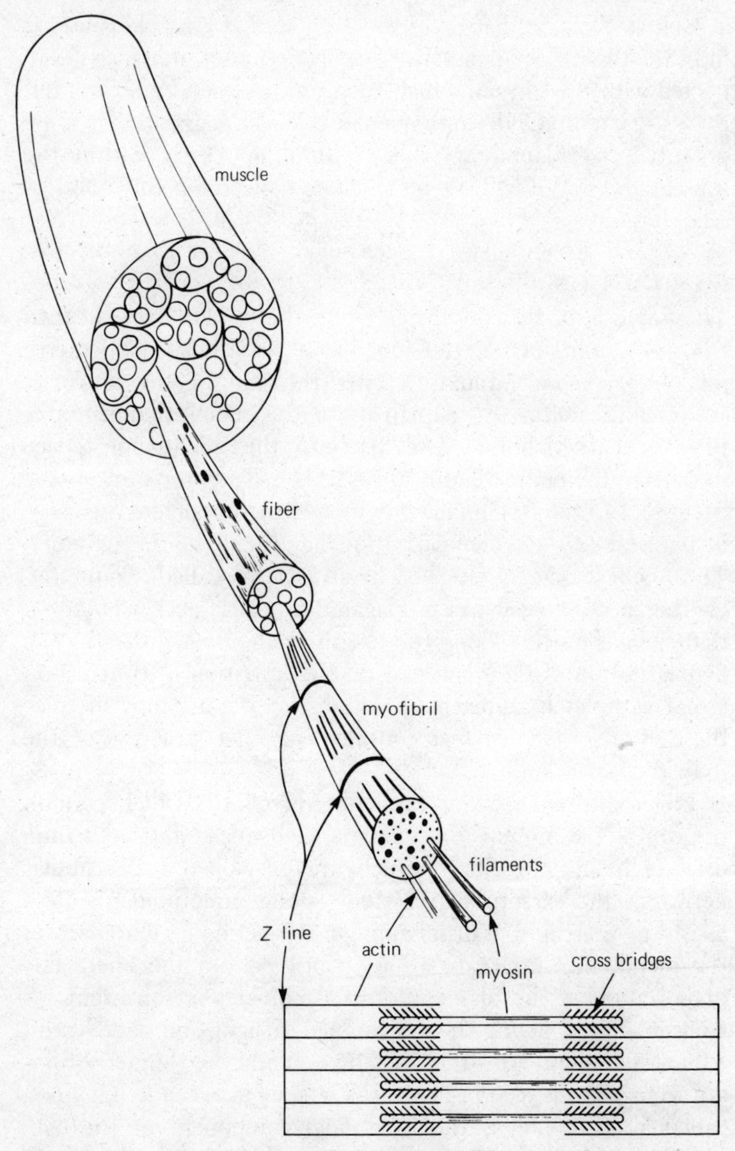

Figure 2.2. The structure of the muscle fiber and the sarcomere.

protein filaments to slide past each other in order for tension to be produced.

More recent electron microscope and X-ray diffraction studies have added considerably to the sliding filament theory of muscular contraction. Cross-bridges have been detected on the thicker myosin filaments. The sliding that shortens the distance between Z lines and produces tension is presumably due to the formation of cross-linkages between the actin (with tropomyosin and troponin) and myosin. As shortening continues, the ends of the actin filaments move toward each other and may possibly overlap. Most of the events of contraction agree with this model. The points of disagreement are small, and the bulk of the data provides firm support for the sliding filament theory. The decrease of tension in the shortening muscle may be explained either by the passage of actin filaments over bridge-free regions of the myosin, or by the overlap of actin filaments, or both. But it is difficult to explain why the stretched muscle produces greater tension than does one at its resting length since stretch diminishes the region of actin and myosin overlap. The answer to this dilemma may be that cross-bridges are stronger at the ends of myosin filaments or that the stretch brings all the sarcomeres into effective position for the transmission of tension to the tendon.

Stimulation of the muscle eventually results in the chemical splitting (hydrolysis) of adenosine triphosphate (ATP) and the release of large amounts of stored energy. It may be that the action potential triggers a release of calcium ions which activate the ATP-splitting ability of the myosin molecules. This ATP-splitting (ATPase) activity probably takes place at the cross-bridges. The myosin cross-bridges swivel to a different angle, pulling the thin filaments past the thick ones. (Murray and Weber, 1974.) These tension-producing cross-linkages may be made and re-formed as successive stimuli reach the muscle. In contractions that allow the shortening of the muscle (isotonic), new bridges will be formed as the filaments creep along each other. In contractions that resist the shortening (isometric), the same bridges may be formed and re-formed after slack is removed from the system.

A substance that binds the calcium ions has been isolated

from muscle tissue samples. This relaxing factor, which seems to be attached to the sarcotubular system, may gather up the calcium ions after a contraction and allow the muscle to return to a relaxed state.

CELLULAR PROCESSES

Before continuing our exploration of the sources and production of energy in the muscle fiber, it is necessary to review some important cellular processes and their relationship to exercise. All cells are encased in a cell membrane, and the muscle fiber is no exception. The cell membrane of the fiber is the sarcolemma, a thin, semipermeable membrane. The sarcolemma exhibits selective permeability such as the alteration in sodium permeability caused by nervous stimulation, and the changing glucose permeability which is evident in cases of diabetes mellitus. Diabetics are unable to utilize glucose despite the large amounts of it present in their blood. Surprisingly, the entry of glucose into skeletal muscle is increased during exercise, even in the absence of insulin. Though the cause of the increased muscle uptake of glucose is not known, it is a vivid and important example of the selective permeability of the sarcolemma. (Ganong, 1971.)

Each muscle fiber (the equivalent of one cell) has many *nuclei* that are found adjacent to the sarcolemma. The nuclei contain the hereditary information encoded on the genes, large molecules of deoxyribonucleic acid (DNA). Though it is unlikely that new muscle fibers can be formed through cell reproduction (mitosis), the cellular code is important in the control of protein synthesis and other cellular functions. Since all enzymes are partly protein and since the contractile filaments are protein, the role of the muscle nuclei cannot be ignored.

The sarcoplasm is the cytoplasm of the muscle fiber. In it are found the various constituents of the cell, including glycogen granules and the enzymes for glycogen metabolism. Other constituents of the sarcoplasm are the mitochondria and ribosomes.

The *mitochondria* are small rodlike structures that appear to have the same function in all cells. The aerobic or oxygen-utilizing enzymes found here provide the major source of ATP.

Mitochondria are more profuse in red endurance muscles and in the heart. The ATP produced by these muscles is essential for many of the energy-using processes of the cell. The mitochondria of muscle cells include the enzymes of the citric acid cycle, fatty acid oxidation, and the electron transport system (Fig. 2.5).

Protein building seems to be the function of the *ribosomes*, the small granular particles usually attached to the sarcotubular system. The ribosomes contain ribonucleic acid (RNA) which aids the process of protein synthesis in two ways. Messenger RNA (*m*RNA) brings the message concerning desired protein from the nucleus. Transfer RNA (*t*RNA) places the appropriate amino acid molecules into the growing protein molecule in the sequence dictated by the RNA template. The energy required for the process comes from ATP molecules manufactured in the mitochondria. Since researchers are only beginning to study mammalian ribosomal preparations, the role of the ribosome in strength (actin and myosin synthesis) and endurance training (aerobic enzyme synthesis) has yet to be explored.

We have already mentioned the presence of *enzymes* in both the cytoplasm and the mitochondria. Enzymes have been defined as organic catalysts that influence chemical reactions without undergoing any permanent changes themselves. Since most of the intracellular reactions related to exercise involve enzymes, it is well that we pause a moment to study their make-up and the factors regulating their activity. All enzymes are proteins, and many also have a nonprotein portion called a *coenzyme*. The protein portion, by virtue of its molecular shape, dictates which compounds the enzyme will influence. The coenzyme seems to be the active portion of the catalyst, the portion necessary to carry out the reaction. Vitamins, trace elements, and pigments have been identified as coenzymes. One of the B vitamins is an essential coenzyme in carbohydrate metabolism.

Enzymes are rather specific in regard to function. Each one usually influences but one type of reaction, and they are often named on that basis (always with *ase* at the end of the name). *Oxidases* add oxygen; *dehydrogenases* remove hydrogen. Some are named for the compound on which they act (*e.g.*, *lipase* acts on lipids). Some enzymes move electrons, atoms, or molecules from one compound to another while others merely move them about on the same compound. Enzyme activity depends upon several factors—temperature, *p*H, substrate availability—and each of these factors can be influenced by exercise.

As the temperature of the cell rises, the enzyme activity increases because of the greater movement of molecules within the cell. Thus the rise in body temperature during exercise can be viewed as a beneficial adaptive mechanism, allowing a more rapid production of ATP. The metabolic implications of warming up, though appreciated by most athletes and coaches, have yet to be fully explained experimentally. However, it seems safe to say that active and related warm-up procedures have the potential to enhance those activities that depend on the continued supply of energy via enzymatic processes. Other reasons for warming up include skill rehearsal, prevention of injury, and perhaps some strength or speed improvement, although the improvements sometimes noted for the last may be related to the removal of inhibition with practice.

Enzymes seem to be affected by the degree of acidity or alkalinity (*p*H) within the cell. Each enzyme seems to perform best at a specific *p*H, and alterations in the *p*H seem to reduce enzyme activity. Exercise of a vigorous nature usually leads to the production of lactic acid. Although some of the acid quickly diffuses out of the cell, the cell itself becomes more acidic (i.e., the *p*H is lowered). It is not clear to what extent this shift in *p*H influences the enzymatic pathways involved in the production of ATP.

The rate of enzyme activity is also influenced by the amount of available substrate and the concentration of enzymes. For example, glycolytic enzyme activity will decrease as the substrate (glycogen) becomes less available, as in an endurance run. Aerobic enzyme activity within the mitochondria depends in part on a supply of pyruvic acid derived from glycogen or glucose metabolism. The supply of enzymes can be influenced by training. Thus, the regular practice of physical activity can influence the future capacity for activity by increasing the concentration of specific enzymes within the muscle cell as well as the amount of available substrates. In summary, enzyme activity is influenced by temperature, *p*H, and the concentration of substrates and enzymes within the cell. The next section shows how these enzymes act on specific substrates to produce energy in the form of ATP.

ENERGY FOR CONTRACTION

Energy—the ability to do work—comes from the sun and is converted into chemical forms by plants and animals, eventually finding its way into your body in the form of carbohydrate, fat, or protein molecules. The oxidation of these molecules provides the energy needed to drive the human machine. However, we have said that the energy for contraction comes from the splitting of high-energy (ATP) molecules. How are the foods we eat converted into high-energy compounds? How does oxidation take place within the cell, and how are we able to exercise briefly beyond our ability to supply oxygen to the muscles?

ATP is the source of energy for most cellular processes. Therefore, the energy in the foods we eat must somehow be released and stored in the form of ATP. ATP is a compound consisting of a complex adenosine molecule and three (tri) phosphate groups.

$$A - P{\sim}P{\sim}P \quad \text{and} \quad A - P{\sim}P + P + \text{energy}$$

The latter two phosphate groups, joined in high-energy bonds that require numerous enzymatically controlled steps to produce, quickly release the stored energy to power the contractile process. The supply of ATP in the muscle cell is not great. Thus, if we are to continue muscular contractions for any length of time, the pathways leading to the production of ATP must be activated. The following list summarizes the major features of the metabolic pathways that provide energy for muscular contraction. (Astrand and Rodahl, 1970.)

Anaerobic
1. ATP→ADP + P + energy for contraction
2. Creatine phosphate + ADP→creatine + ATP
3. Glycogen or glucose + P + ADP→lactate + ATP

Aerobic 4. Glycogen and free fatty acids + P + ADP + $O_2 \rightarrow CO_2 + H_2O$ + ATP

Phosphagen

Creatine phosphate (CP) is another high-energy compound that provides the energy to resynthesize ATP during contractions. However, like ATP its supply is also limited. During exercise both the ATP and CP supplies in muscle are depleted, with by far the greater decline in CP (a 25-percent decline for ATP versus an 85-percent drop for CP during exhaustive work; Karlsson and Saltin, 1970). This conservation of ATP during exhaustive effort has been interpreted as evidence that some muscle ATP is stored elsewhere and is unavailable to the contractile apparatus. This is consistent with the idea that ATP is needed for cellular processes other than contraction.

The combined ATP and CP have been labeled as phosphagen (ATP + CP = phosphagen), or the high-energy phosphate stores. (Margaria et al., 1966.) This portion of anaerobic metabolism can only continue at peak output for 4–5 seconds. However, it is this component of anaerobic energy (steps 1 and 2 above) that powers the explosive charge of the lineman in football, the sprinter from the blocks, and the other maximal bursts of effort that add such excitement to sport. Margaria and his colleagues have suggested a simple, all-out stair-run test to determine the maximal anaerobic power. Using the time required to propel the body through the vertical distance of several stair steps we can easily compute power in foot pounds per second. The procedure would seem to hold some pedagogical and motivational value for the teacher or coach interested in measuring an important phase of anaerobic metabolism (see Table 9.1).

Lactate

The next step in the anaerobic scheme (step 3) involves the breakdown of muscle glycogen or glucose (sugar) to provide the energy for ATP production. This process—glycolysis—leads to the formation of two ATP molecules and lactic acid, in the absence of oxygen. The anaerobic metabolism is that portion of energy production that goes on when the oxygen supply is insufficient to meet the demands. While the intense bursts of effort supported by phosphagen stores can only be sustained at peak levels for about 4–5 seconds, total anaerobic effort (phosphagen plus lactate) can continue for 40–50 seconds. This does not mean that athletes collapse after 50 seconds of maximal effort, although some do after a race like the 440-yard dash. It only means that when they have exhausted the anaerobic potential the intensity of activity must be reduced to allow slower aerobic processes to meet the energy needs. Therefore, the total anaerobic system is essential to the nature of sport as it is practiced in our society. Many of our most popular athletic contests rely heavily on anaerobic processes. Consequently, it is not surprising that researchers have attempted to determine the power or capacity of the anaerobic system in order to better study the effects of training and the relationship of anaerobic training to performance in sport.

Anaerobic metabolism has typically been studied by measuring the blood lactate concentration or by measuring the oxygen debt capacity. Lactate produced during vigorous exercise diffuses from the muscle fiber into the surrounding tissues and eventually reaches the circulation (Fig. 2.3). From there it can be taken up by the heart, liver, or kidneys, and removed from the circulation. It is not entirely clear to what extent blood lactate levels parallel lactate production. Cardiac muscle oxidizes lactate, the liver converts it to glycogen, and, after exercise, the kidneys may remove remaining lactate from the circulation. While blood lactate has been used as a gross indicator of anaerobic processes in the same subjects performing the same tests, it cannot be said with certainty whether the increased levels of blood lactate reflect increased anaerobic capabilities or a greater ability to tolerate a lactate accumulation (Cunningham and Faulkner, 1969). Recent evidence suggests that some of the pyruvic acid produced in anaerobic glycolysis may be converted to the amino acid *alanine* instead of the expected lactic acid. The alanine is released to the circulation, taken up by the liver and converted to glucose, thus completing the glucose-alanine cycle. One effect of training may be an increase in this alternative pathway for pyruvate metabolism, with a corresponding decrease in the production of lactic acid

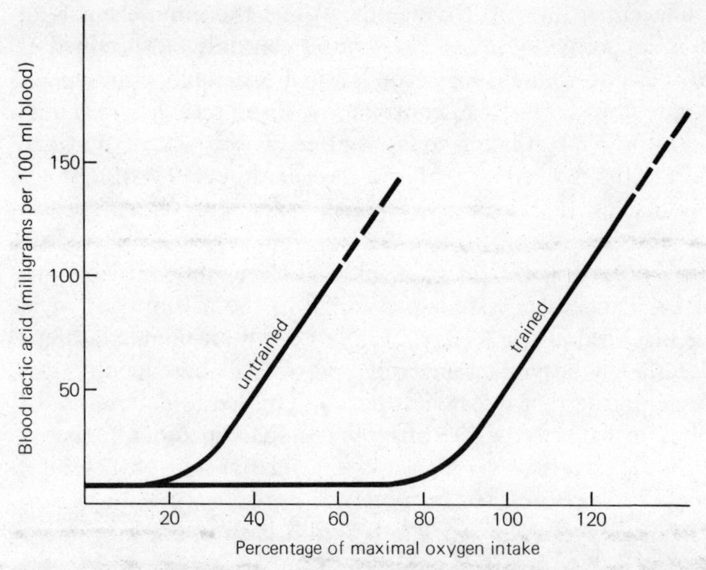

Figure 2.3. Exercise intensity and the production of lactic acid. Unfit or untrained individuals begin to accumulate lactic acid at a low percentage of their maximal oxygen intake (aerobic capacity). Trained individuals are able to work longer at a higher percentage of maximal oxygen intake without the accumulation of lactic acid. Thus, training increases the maximal oxygen intake and allows the individual to persist at a higher percentage of that maximal intake.

at a given work load. (Molé, Baldwin, Terjung, and Holloszy, 1973.) These recent findings will undoubtedly require a reassessment of the classic interpretations of lactic acid measurements.

Another method used to indicate the extent of anaerobic metabolism employs the collection of expired air samples after exercise has been terminated. The amount of oxygen consumed during the recovery period exceeds the oxygen needs at rest and, therefore, has been termed the *oxygen debt* (Fig. 2.4). It represents the oxygen required to pay off the debt incurred during anaerobic exertion. The greater the intensity of exercise, the greater the debt. Researchers have identified two components of the oxygen debt repayment process. The rapid initial phase, called the alactacid oxygen debt (nonlactate portion), has been associated with the replenishment of phosphagen stores. The slower *lactacid* portion of the debt is utilized to replace the liver and muscle glycogen that was depleted during the contractions. The lactacid portion is far slower since the conversion from lactic acid to glucose to glycogen (long chain of glucose molecules) takes place in the liver. Muscle glycogen stores are replenished from glucose transported from the liver via the circulation (the Cori cycle). Thus, this classic view of the oxygen debt would seem to afford a method for the measurement of anaerobic metabolism.

However, recent research has suggested a different interpretation of the oxygen debt that tends to diminish its usefulness

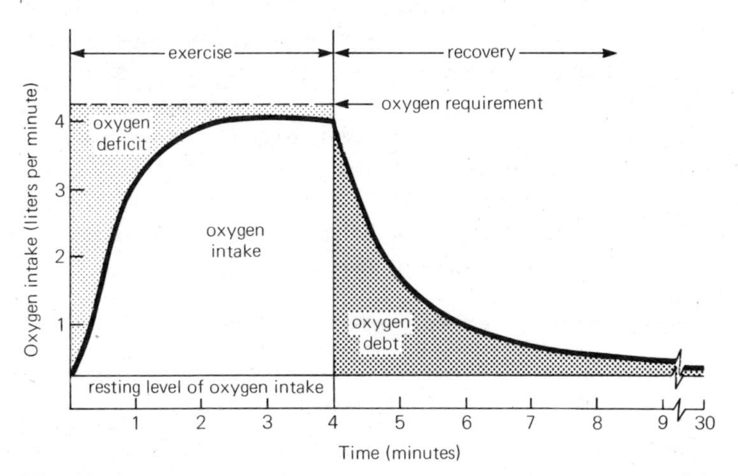

Figure 2.4. Oxygen intake, oxygen deficit, and oxygen debt. Oxygen intake does not adjust immediately to the demands of exercise so a deficit is incurred in the early moments of exertion. When the oxygen intake capacity (4 liters in this example) does not meet the requirements of the exercise a further deficit is encountered. Since the requirement exceeds the intake capacity the debt must be repaid during recovery. The rapid initial portion of the recovery curve may serve to repay the alactacid debt associated with phosphagen depletion at the beginning of exercise. The slower portion of the debt repayment serves to remove the lactic acid formed during anaerobic glycolysis. The debt may exceed the deficit when the requirement exceeds the oxygen intake capacity (aerobic capacity).

as an indicator of anaerobic metabolism. In addition to its importance for the resynthesis of phosphagen, particularly the severely depleted CP, the debt also serves to return oxygen to the muscle myoglobin and to the blood to restore resting levels of oxygen saturation. Furthermore, one always wonders how long to continue to collect the oxygen debt and whether the oxygen required at rest before exercise is the same required at rest during recovery.

At present, there does not seem to be any simple reliable estimate of anaerobic capacity or power that can be utilized to measure the effects of training on anaerobic potential. We have found that all-out exercise bouts can be improved through training designed to tax the anaerobic mechanisms. Using short-interval sprints, Cunningham and Faulkner (1969) and Dainty (1971) have found increased performance capacities as measured by short, exhaustive exercise tests on the treadmill and bicycle ergometer respectively. Though the increased exercise times seemed to relate to increased oxygen debt capacities, it was not clear to what extent aerobic capabilities influenced the performance changes (Table 2.1). In Chapter 3 we will see how researchers have examined the muscle fiber to determine the effects of training on both anaerobic and aerobic pathways.

TABLE 2.1. OXYGEN DEBT COMPARISON

	Oxygen debt (liters)	
Subjects	*Men*	*Women*
Athletes	10–18	6.0
Active	6–9	4.0
Inactive	< 6.0	3.0

Sources: Hermansen, 1969; Dainty, 1971.

Aerobic Pathways

As indicated in Fig. 2.5, the aerobic portion of energy production utilizes glycogen and free fatty acids (FFA) as well as oxygen. The aerobic oxidation of a molecule of glucose yields 38 molecules of ATP. Contrast that with the 2 molecules of

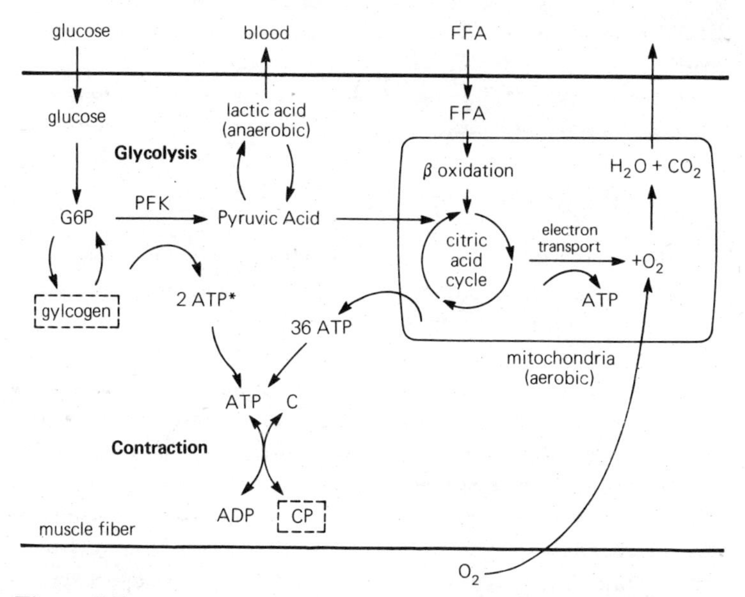

Figure 2.5. Metabolic pathways and the production of adenosine triphosphate (ATP). Limited glycogen and creatine phosphate (CP) supplies are depleted at the start of exercise, during high-intensity effort, and during the transition to muscular exhaustion. Blood glucose is spared for use by the central nervous system. Free fatty acids (FFA) provide the major fuel for long-term, steady-state contractions. (*Anaerobic: Glucose → pyruvate = gain of 2 ATP. Aerobic: Glucose → CO_2 + H_2O = 38 ATP.)

ATP produced in the anaerobic metabolism of glucose to lactic acid. It is clear why we are able to continue almost indefinitely when operating aerobically, and why any exercise that leads to lactate production will also reduce the time in which the exercise can be continued at such a rate. When the exercise intensity is relatively light and the oxygen supply is equal to the demand, glycolysis proceeds to form pyruvic acid, which then enters the Krebs cycle (citric acid cycle) located in the mitochondria. The enzymatic steps of the Krebs cycle release the energy bound in the chemical bonds, and the entering compounds are reduced to carbon dioxide and hydrogen atoms. But this is not the pathway that utilizes the oxygen, nor is it the cycle that generates most of the aerobically produced ATP.

Electrons from the hydrogen atoms are passed along the electron transport system. The energy released in the passage is used to form ATP. The *final step* in the process is the combination of hydrogen and oxygen to form water. So, for all the steps involved in aerobic or oxidative metabolism, only the final ones involve oxygen, in an elegantly controlled internal combustion system.

Fatty acid molecules are liberated from adipose tissue and carried in the circulation to the working muscles. They enter the mitochondria where they are fragmented in a process called *beta oxidation.* Two carbon fragments enter the Krebs cycle for further oxidation to carbon dioxide and water. During rest and moderate exercise, fat forms the major source of energy for muscle cells. When exercise intensity increases, the energy-rich but oxygen-poor fat molecules become a burden to the already strained oxygen transport system. Thus the muscles shift to glycogen as an energy source. If exercise intensity becomes even greater and the oxygen supply cannot keep pace, glycogen metabolism will result in the formation of lactic acid and the phosphagen stores will become depleted.

Exercise intensity determines the metabolic pathway utilized. But *intensity* is a relative term which depends upon each individual's exercise capacity. Exercise of low intensity for the athlete may be maximal for a sedentary individual. The *aerobic capacity* or *maximal oxygen intake* defines the maximal rate at which we can consume oxygen. Ultimately, it depends upon the capacity and efficiency of the aerobic pathways. Therefore, the individual with a high aerobic capacity can operate aerobically while one of lesser capacity must dip into his lactate or phosphagen reserves. The fit individual (i.e., one having a high aerobic capacity) utilizes fat as an energy source while the unfit must rely on limited carbohydrate supplies. In Chapter 3 we will discuss how diet and training influence the fuel for contraction and the capacity for exercise as well. At this point, it should be noted that the aerobic capacity *can* be measured directly or be predicted by using maximal or submaximal testing procedures (on the treadmill, laboratory bicycle ergometer, or step test; see Fig. 2.6).

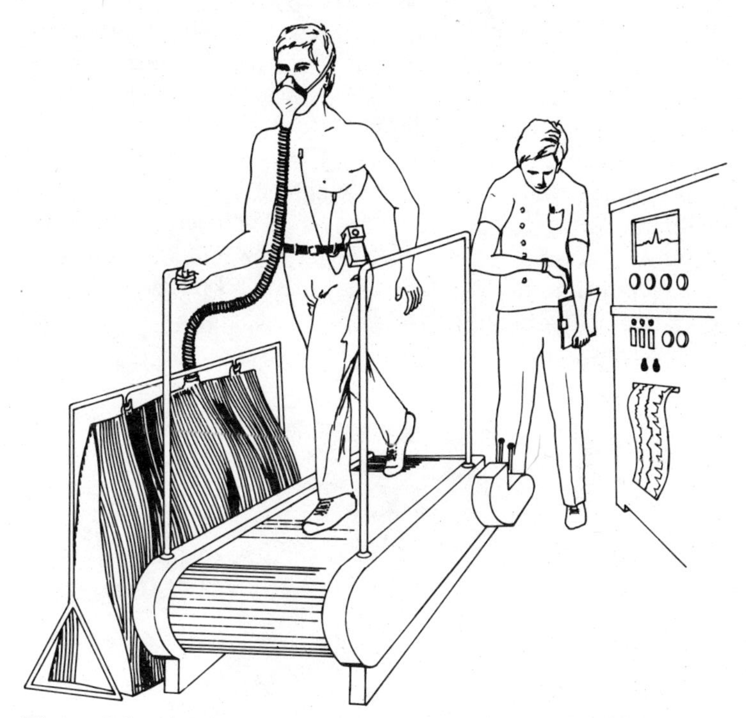

Figure 2.6. A laboratory ergometer known as the "treadmill."

Studies have shown tremendous individual differences in aerobic capacity. Distance runners and cross-country skiers are able to consume oxygen at twice the rate possible for inactive individuals (almost 80 ml per kilogram of body weight per minute versus 40 ml for the sedentary). The aerobic capacity serves as the most important indicator of physical fitness because it involves the ability to take in (respiration), transport (circulation), and utilize oxygen (aerobic enzymes) in the active muscles. More will be said about this important measure in later chapters.

The use of the anaerobic or aerobic pathways depends on a number of factors including the intensity of the exercise, the fitness of the individual (aerobic capacity), his experience or efficiency in the exercise, and the length of the event. High-intensity events of short duration rely on anaerobic processes, while long and necessarily less intense activities utilize the extensive aerobic energy stores. However, we do not shift abruptly from one energy source to another. On the contrary, the transition from aerobic to anaerobic processes in exercise of increasing intensity is rather subtle. In Chapter 3 we shall see how these processes overlap and how they are influenced by diet and training. In the process we shall extract some of the principles of exercise and training that are essential to the intelligent conduct of physical activity and sport programs.

CHAPTER 3 DIET, TRAINING, AND THE FUEL FOR CONTRACTIONS

Let us begin this discussion with a brief look at the principal fuels—carbohydrate and fat—noting how they are used during rest and exercise. We will omit a discussion of protein as a fuel since its typical contribution is quite small. Protein only serves as a significant source of energy during periods of starvation.

FUELS FOR CONTRACTION

Fat is a far more efficient way for us to store energy because it contains more than twice the energy per gram than does carbohydrate (Table 3.1). Furthermore, since the average man carries about 10–15 percent of his body weight in fat and but a small fraction of his

TABLE 3.1. CALORIC EQUIVALENTS OF FOODS[a]

Food	*Energy (calories per gm)*	*Oxygen Required (liters per gm)*
Fat	9.3	1.98
Carbohydrate	4.1	0.81
Protein	4.3	0.97

[a] Calories (cal) refer to kilocalories (kcal) or the amount of heat energy required to raise the temperature of 1 kg of water 1 C.

weight in carbohydrate, he will have fifty times more energy stored as fat. On the other hand, since fat is rather poor in oxygen it requires more oxygen to burn a gram of fat. When we compare fat and carbohydrate in terms of energy per liter of oxygen used, we find that carbohydrate is somewhat more efficient as a fuel. Therefore, it is not surprising that we shift from fat to carbohydrate for energy when exercise intensity increases.

Fat consists of a molecule of glycerol and three molecules of a fatty acid (Fig. 3.1), ranging from short carbon chains such as butyric acid (C_3H_7COOH) to longer chains containing sixteen or more carbon atoms ($C_{15}H_{31}COOH$, or palmitic acid). Upon digestion, fat is split into fatty acids and glycerol, and absorbed into the lymphatic system. From there is passes to the blood where it may be transported for use as a fuel or for deposit in adipose tissue. During exercise, epinephrine stimulates the release of free fatty acids (FFA) from adipose tissue storage. Carbohydrates enter the bloodstream and are carried to the liver. There they replenish the liver glycogen supply, maintain blood sugar levels, and eventually replace glycogen depleted from muscle fibers.

The respiratory exchange ratio (R) or the ratio of the volume of carbon dioxide expired per minute ($\dot{V}CO_2$) to the volume of oxygen consumed ($\dot{V}O_2$) serves as an indication of the food being metabolized (CO_2/O_2).

Palmitic acid + $23O_2 \rightarrow 16CO_2 + H_2O$ (16/23) $R = 0.7$
Glucose + $6O_2 \rightarrow 6CO_2 + H_2O$ (6/6) $R = 1.0$

CO_2 Produced (liters per gm)	Respiratory Exchange Ratio ($R = CO_2$ per O_2)	Caloric Equivalent (calories per liter of O_2)
1.40	0.7	4.696
0.81	1.0	5.061
0.78	0.8	4.432

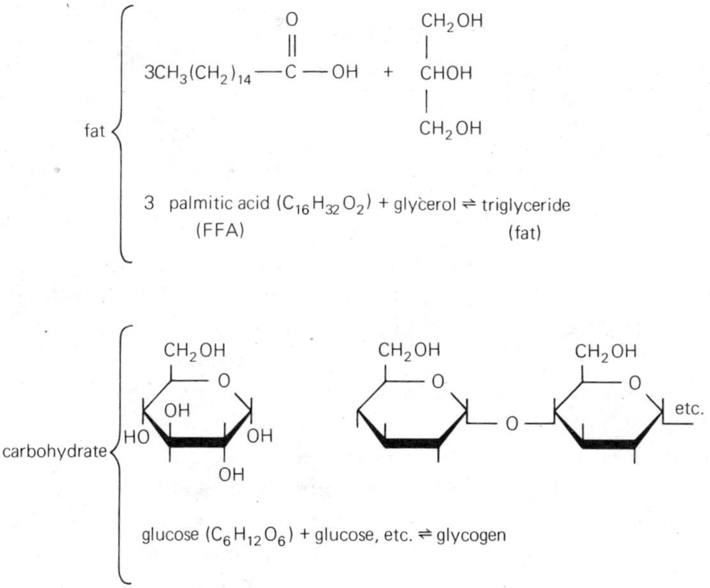

fat

$$3CH_3(CH_2)_{14}{-}\overset{O}{\overset{\|}{C}}{-}OH \; + \; \begin{array}{c} CH_2OH \\ | \\ CHOH \\ | \\ CH_2OH \end{array}$$

3 palmitic acid ($C_{16}H_{32}O_2$) + glycerol ⇌ triglyceride
(FFA) (fat)

carbohydrate

CH₂OH O OH HO OH OH — CH₂OH O — O — CH₂OH O etc.

glucose ($C_6H_{12}O_6$) + glucose, etc. ⇌ glycogen

Figure 3.1. Fat and carbohydrate molecules. Note the relative absence of oxygen in the fatty acid formula. Triglycerides and glycogen are the storage forms of fat and carbohydrate respectively. The fat is stored in adipose tissue and the carbohydrate is stored in the liver and muscles. A small amount of fat is also found in the muscles.

R usually ranges from about 0.8 at rest to 1.0 during vigorous exercise. This suggests the metabolism of fat and carbohydrate at rest, and a shift toward carbohydrate metabolism as exercise intensity increases. As indicated in Table 3.1, the energy (caloric) equivalent of fat and carbohydrate range from 4.696 to 5.061 kcal per liter of oxygen respectively. For the sake of simplicity, we shall adopt the value of 5 kcal per liter of oxygen as an average for future considerations. Thus we burn about 5 calories (kcal) for each liter of oxygen we consume.

Since it is difficult to directly measure the heat generated by an exercising subject, exercise physiologists have adopted an indirect method for the measurement of energy expenditure. The popular *open circuit* method involves the collection of expired air and the analysis of its contents to determine the amount of oxygen extracted from the atmosphere. Thus, if a subject exhales air of 16.9 percent oxygen and the atmosphere contains 20.9 percent, the difference of 4 percent has been utilized in energy metabolism. By multiplying this by the amount of inspired air (e.g., 50 liters per minute), we are able to calculate the oxygen consumption (4 percent $\times$ 50 $=$ 2 liters of oxygen per minute). The net cost of the activity can then be determined by subtracting the resting oxygen consumption (e.g., 2 liters $-$ 0.250 $=$ 1.75 liters of oxygen per minute). The energy cost in calories can then be approximated by multiplying the average of 5 kcal per liter against the net cost (1.75 liters). The energy cost calculated in our example (8.75 calories) would be equivalent to that required for easy jogging or a vigorous game of tennis (see Appendix C).

From Rest to Exercise to Exhaustion

It is clear that aerobic pathways must be utilized to the maximum if we are to delay the onset of fatigue. Also, it seems important to preserve extramuscular (blood) glucose supplies for use by tissues that depend on glucose as an energy source. It is not surprising, therefore, to see that the resting muscles consume fat (FFA) almost exclusively. During the transition to physical activity, when the circulation and respiration have yet to adjust to the demands of the task, the energy for replenishing adenosine triphosphate (ATP) comes from the available creatine phosphate (CP) pool as well as from the anaerobic breakdown of muscle glycogen. If prolonged submaximal activity is studied, blood glucose may also be seen to contribute to the regeneration of the high-energy phosphate compounds during the transitional state. When a steady state is reached and ATP resynthesis is equal to ATP breakdown, the energy for continued activity comes from the oxidation of FFA, glycogen, and blood glucose. The intensity of the activity dictates the ratio of FFA to glycogen oxidation, with glycogen utilization increasing as the work intensity increases. As we shall soon see, diet and training also influence that ratio.

When the exercise intensity leads to exhaustion within 10 minutes, the CP is severely depleted, but only a portion of the muscle glycogen has been used. When exhaustion follows prolonged strenuous work, the muscle glycogen is depleted along with the CP (Table 3.2). Thus exhaustion occurs when the glycolytic and oxidative enzymes are unable to keep up with the demands for ATP (Weiser, 1971), and when the appropriate sources of energy have been depleted.

Several key factors influence the pathways utilized to produce energy for contractions. The title of this chapter suggests the importance of diet and training in that regard. Another factor—the intensity of exercise—deserves attention at this point, since it is a significant determinant of the pathway utilized and the substrate chosen.

Intensity of Exercise

The exercise intensity can be defined by using the oxygen (or caloric) consumption per unit of time. As the speed of running increases, the cost of running increases. Thus, it takes about 2 liters of oxygen per minute to run a 12-minute mile (5 mph) and about 4 liters of oxygen per minute to run a 6-minute mile (10 mph). The higher the intensity the greater is the oxygen deficit. Since the deficit is met with limited glycogen and CP energy stores, a large deficit will limit the time the exercise can be continued at that intensity. The intensity also dictates the ratio of FFA to glycogen utilized during the steady state. The higher-intensity exercise requires a shift from the oxygen-poor FFA to glycogen or glucose, a move that could eventually

TABLE 3.2. MAIN SOURCES OF ENERGY FOR MUSCULAR CONTRACTIONS[a]

Rest	FFA
Exercise begins	CP + glycogen
Steady state	FFA + glycogen[b] (↗ intensity = ↗ glycogen)
Exhaustion	CP + glycogen (short intense = CP ↘; prolonged = glycogen ↘)

[a] Abbreviations: FFA, free fatty acids; CP, creatine phosphate.
[b] High blood glucose levels tend to supplement FFA + glycogen.

reduce the blood sugar and reduce the essential glucose supply to the nervous system. The higher-intensity exercise would hasten the reliance on glycogen and CP stores, and thereby lead to exhaustion. Finally, high-intensity effort requires anaerobic glycolysis and leads to the production of lactic acid.

The exercise intensity can best be expressed as a percentage of one's maximal oxygen intake. The lower intensity corresponds to a lower percentage of maximal oxygen intake, a smaller oxygen deficit, increased FFA utilization, and an increased work time. The higher the aerobic capacity, the greater the body's ability to utilize aerobic pathways in a given work task, such as a running event in track. Since each of us differs in our aerobic and anaerobic capacities we also differ somewhat in how we pay the oxygen costs of various running events. Those with high aerobic abilities (a high percentage of slow-twitch fibers) would feel more comfortable in the longer, slower events, and of these, those with higher anaerobic capabilities (fast-twitch fibers) might choose to utilize a finishing kick while others might prefer a steady pace throughout.

When the exercise intensity exceeds the aerobic capacity we call upon less efficient anaerobic pathways to supply ATP. In such cases we are limited by the combined aerobic and anaerobic capacities. How long could a man with maximal oxygen intake of 4 liters and a maximal oxygen debt of 16 liters continue to run at a pace of 12 mph (a 5-minute mile) costing 5 liters of oxygen per minute? Theoretically, were it not for the oxygen deficit at the start of the exercise, he would be going into debt at the rate of about 1 liter per minute and could continue for 16 minutes. Theoretically, a distance runner with an aerobic capacity of 5 liters could go on for a considerable distance in spite of his smaller anaerobic capacity. World class marathon runners are now averaging 5 minutes per mile while running the 26-mile, 385-yard distance. However, they also utilize anaerobic pathways since no one is able to continue for very long at 100 percent of his aerobic capacity.

TRAINING AND THE FUEL FOR CONTRACTIONS

This section outlines the effects of training on the metabolic pathways and, consequently, on the fuels used for contractions. We will first consider the phosphagen and lactate portions of anaerobic metabolism, and then dwell on the aerobic energy pathways. Most of the findings considered here have been published since 1967. Included are some new techniques that have revolutionized the conduct of exercise physiology research. The needle-biopsy technique introduced by Bergström (1962) has become an essential tool for the analysis of the effect of training and nutrition on muscle tissue. It has allowed the repeated analysis of human muscle samples. Now the results of animal studies can be correlated with corresponding data from humans, performance data can be related to tissue changes, and theories can be tested at the cellular level.

Phosphagen

Since the depletion of CP marks the transition to exhaustion in high-intensity effort, we would expect an increase in this high-energy compound to extend anaerobic capabilities. Some time ago Palladin (1945) reported elevated CP levels due to training. More recently, Gale (1970) investigated the effects of several types of training on the biochemical and histochemical properties of rat muscle. Rats trained in an interval-sprint program experienced increased levels of CP. Earlier, we had noted that the CP was severely depleted in short-term exhaustive exercises, such as interval sprints. Therefore, it seems likely that significant depletion of the CP stores during short-term exhaustive exercise serves as a *training stimulus* for the adaptive increases noted.

Lactate

The enzymes of the glycolytic pathway have been studied with reference to the effects of training. Although it is known that training increases the capacity to produce or tolerate high lactate levels, most investigations conclude that training does not alter the glycolytic enzymes. This may have been due to the nature of the training programs used or to the selection of an inappropriate enzyme for study. Recent study of the rate-limiting enzyme phosphofructokinase (PFK) has revealed a training effect caused by a strenuous training program (Gollnick et al., 1973). Earlier Gollnick (1971) had suggested that improvement of the glycolytic system may be unnecessary, especially since the pathway normally produces an excess of substrate for the oxidative processes.

Glycogen, the major substrate of the glycolytic pathway during the early and latter stages of prolonged exercise also seems to be subject to a training effect under specific conditions. We shall explore these conditions in the section entitled, "Diet and the Fuel for Contraction."

Recent studies suggest that training alters the anaerobic threshold (AT), the point at which lactic acid begins to appear in the blood (see Fig. 2.3). While unfit individuals may begin to produce lactate (via anaerobic glycolysis) at work loads equivalent to 20 percent of their maximal oxygen intake capacity, fit individuals are able to work at levels beyond 80 percent of maximal effort before lactate appears in the blood (Londree, 1973). This suggests that the fit individuals are able to utilize more efficient aerobic pathways over a wider range of work loads.

Unfit:
20 percent $\times$ 40 ml/kg/min = 8 ml/kg/min (AT)
Fit:
80 percent $\times$ 80 ml/kg/min = 64 ml/kg/min (AT)

No wonder highly fit endurance athletes can continue for hours at work loads that would exhaust the unfit in minutes.

In summary, although the effects of anaerobic training are incompletely understood, it does seem that training enhances the capacity of the system and the performances the system supports. Additional effects of anaerobic training on athletic performance may be due to the effects of the training on *aerobic capacity* and the resetting of the anaerobic threshold.

Aerobic Pathways

The influence of training on the aerobic fuels and pathways was poorly understood prior to 1967. It was known that endurance training improved performance in distance running, swimming, cycling, and skiing. However, the magnitude of the improvement could not be explained by the changes in the cardiovascular or respiratory systems. Thus researchers were interested in possible cellular effects, despite the fact that earlier efforts had failed to uncover significant changes in the aerobic pathways.

Holloszy (1967) reasoned that earlier training studies had failed to adequately load the aerobic pathways. Consequently, he subjected rats to a very strenuous running program on the treadmill. Trained rats were able to continue exercise for 4–8 hours while control animals became exhausted after 30 minutes. Following training, the rats were sacrificed and muscle mitochondrial fractions were prepared via centrifugation. These samples were then tested for their ability to oxidize pyruvate, the end product of glycolysis and the starting point of aerobic metabolism in the Krebs cycle. The oxygen uptake of the mitochondrial samples from trained animals was twice that of the sedentary controls. A 50–60 percent increase in mitochondrial protein (per gram of muscle) was also noted. Specific enzymes in the Krebs cycle and the electron transport system doubled in oxidative capacity. These findings, along with the tightly coupled oxidative phosphorylation exhibited by the trained animals indicated that the doubling of mitochondrial electron transport capacity was associated with a similar increase in the ability to generate ATP aerobically. (Holloszy et al., 1971.)

To further investigate the effect of training on aerobic pathways, Gollnick and King (1969) also subjected rats to a vigorous training program. Upon sacrifice, muscle samples were prepared for electron microscopic study of the ultrastructure. The mitochondria of trained rats were increased in size and number, and their cristae appeared to be densely packed (Fig. 3.2). It now appears that the biochemical alterations noted by Holloszy parallel the ultrastructural alterations reported by Gollnick and King. And training seems capable of increasing the oxidative capacity of white and red (FT and ST) muscle fibers. (Baldwin et al., 1972.) Endurance training, by increasing the oxidative or aerobic capacity of the individual fibers, involves a greater proportion of the muscle in endurance activities. This could also allow individual motor units a chance to rest while other units carried out the submaximal work load.

But how do these fascinating changes influence the choice of fuel to be metabolized? How do they function to conserve the limited muscle glycogen stores at the expense of the abundant fat reservoir? Because FFA are poor in oxygen, their use is limited by the oxidative capacity. Poorly understood cellular controls shift to glycogen usage when exercise intensity increases beyond 60–70 percent of maximal oxygen intake. Any increase in the oxidative capacity should prolong the ability to oxidize FFA. Studies have shown that trained animals (and men) are capable of extracting a greater percentage of their energy from FFA during submaximal exercise. (Issekutz et al., 1965.) Convincing proof of the effect of training on FFA oxidation was provided by Molé, Oscai, and Holloszy (1971). The ability of rat gastrocnemius muscle to oxidize the FFA palmitate was significantly increased following a program of treadmill running.

We now have a picture of exercise and training that is beginning to come into focus. While we are still uncertain about some of the limiting factors in certain types of exercise, we see a hierarchy of energy stores and preferred pathways seemingly designed to conserve needed glucose and glycogen. The aerobic metabolism of abundant FFA stores is preferred at rest and for light-to-moderate exercise. Training enhances the utilization of FFA. Anaerobic energy sources in the form of phosphagen are used somewhat at the onset of light-to-moderate work. At a

Figure 3.2. The effects of training on skeletal muscle mitochondria. These illustrations show longitudinal sections of control (*A*) and trained (*B*) rat skeletal muscle. Note that the muscle from the trained animal has larger and more numerous mitochondria and their internal structure or cristae seem more densely packed. Observe also the ultrastructure of the sarcomeres, the myofibrils and filaments. Notice the shaded areas where the actin and myosin filaments overlap. × 11,000.

Source: Gollnick and King, *American Journal of Physiology*, 216 (June 1969): 1502.

moderate work intensity, phosphagen and anaerobic glycolysis are involved along with the aerobic pathways. At heavy work loads, phosphagen depletion and anaerobic glycolysis are at a maximum. (Karlsson and Saltin, 1971.) Training increases CP and glycogen stores and enhances their ability to mobilize and utilize FFA (Table 3.3).

DIET AND THE FUELS FOR CONTRACTION

We have seen that the choice of fuel for muscular contractions depends in part upon the intensity of exercise. The composition of the diet has also been known to influence that choice. In 1939, Christensen and Hansen reported remarkable increases in endurance performance for subjects on an extremely high carbohydrate diet. More recently, the elegant muscle-biopsy technique introduced by Bergström (1962) has been used to study the influence of exercise and diet on muscle glycogen stores and endurance performance. It has been amply demonstrated that several conditions are necessary in order to receive maximal performance benefits.

TABLE 3.3. TOTAL ENERGY AVAILABLE FOR MUSCULAR CONTRACTIONS (IN A 75-kg MAN WITH 28 kg OF MUSCLE[a])

Source	*Supply*	*Energy (calories per mole)*	*Total Energy (calories)*
ATP	4.0 mmoles per kg of muscle	10	0.8
CP[b]	15.6 mmoles per kg of muscle	10.5	3.3
Glycogen[b]	15.0 gm per kg of muscle	700	1,200
FFA	>10% of body weight	VARIES	50,000

[a] Abbreviations: ATP, adenosine triphosphate; CP, creatine phosphate.
[b] Subject to improvement by training.

Sources: Karlsson and Saltin, 1971; Astrand and Rodahl, 1970.

First, the muscle must be *depleted* of its original glycogen stores through prolonged strenuous exercise. Second, the subject is maintained on a low carbohydrate diet for several days of continuing exercise. Following this, the athlete reduces his work load and is placed on a high carbohydrate diet for several days preceding competition. Muscle glycogen stores have thus been elevated from a mixed diet level of 17.7 gm per kilogram of wet muscle to 35.2 gm and in one subject to 64.8 gm per kilogram of wet muscle. (Karlsson and Saltin, 1971.) The best endurance performances were always attained while on the high carbohydrate diet. When the glycogen stores were reduced, prolonged effort could still be sustained at *lower* work intensities using FFA as fuel. (Pernow and Saltin, 1971.) It appears that at high submaximal loads the amount of muscle glycogen determines the long-term performance capacity.

The effect of training and diet on muscle glycogen—*glycogen supercompensation*—allows us to introduce a principle of athletic training that will receive considerable attention throughout this monograph. This principle—*specificity of training*—is illustrated by the fact that supercompensation only takes place in the exercised muscle fibers. Furthermore, because muscle glycogen cannot be removed from one fiber for use in another, training must be specific so as to prepare precisely those muscle fibers to be utilized in competition. The presence of the enzyme glucose-6-phosphatase would be necessary to change glycogen to glucose for transport from one fiber to another. The lack of that enzyme accounts for the specificity of the glycogen supercompensation effect on performance.

Glycogen supercompensation does not permit an athlete to run faster, but it does seem to enhance the maintenance of an optimal pace from the beginning to the end of the race. An explanation for the beneficial effect of glycogen on race-pace is based on the selective use of glycogen at higher exercise intensities. When muscle glycogen is depleted the muscle must rely on fat as an energy source. Since fat requires more oxygen to produce ATP and since the oxygen supply is already limited, the pace is reduced to the level consistent with the oxidative metabolism of FFA.

We should not leave this subject without a word of warning

concerning high carbohydrate diets. There does not seem to be any need for prolonged maintenance of a high carbohydrate diet. You can receive its benefits in about 4–6 days. Excessive glycogen storage has been discouraged due to the fact that glycogen is always stored with water. The increase in body weight from storage of hydrated glycogen could have a negative effect on performance. (Karlsson and Saltin, 1971.) On the other hand, the water could become useful during prolonged work in the heat.

The availability of high concentrations of blood glucose while one is on a high carbohydrate diet seems to increase muscle utilization of glucose. It would not seem wise to condition the metabolic pathways to a high carbohydrate utilization. A shift away from FFA during rest and light exercise could lead to periodic hypoglycemia or low blood sugar and symptoms of CNS fatigue. So the glycogen supercompensation phenomenon should be viewed as an established performance aid, achieved (occasionally) by following the depletion of existing glycogen stores with a low carbohydrate diet and exercise, in turn followed by a high carbohydrate diet and relative inactivity for several days prior to competition.

We have reviewed the effects of training on aerobic and anaerobic metabolism, and examined the influence of a high carbohydrate diet on performance capacity. We are beginning to see that training programs are specific in regard to exercise intensities and the alterations in pathways and substrates. We are also learning that all the fibers of a motor unit are affected similarly, and that usage or training can alter the biochemical nature of the fibers. But we have yet to consider that highly valued property of muscle—strength. In Chapter 4, we shall consider the cellular effects of strength training and relate them to endurance and fatigue.

CHAPTER 4 STRENGTH, ENDURANCE, AND FATIGUE

The tension or force developed in a contraction is the result of the frequency of motor neuron discharge (from 5–50 impulses per second) as well as of the number of motor units activated. Endurance—the repetition of submaximal contractions—can be accomplished either by the same motor units or by a trading off of responsibilities by many units. A maximal contraction would presumably involve high-frequency discharge to all the motor units in an expression of strength. Fatigue can be defined as the inability to continue the contractions. Of interest is the interaction of nervous and cellular events in the phenomena under discussion. Strength, and even the training of strength, seems to involve both nervous and muscle fiber components. Let us begin with a discussion of strength and see how the components combine to allow a most interesting facet of human performance.

STRENGTH

In research, we typically speak of *maximal voluntary strength* since it is probable that the muscle is capable of exerting more force than we can call upon under ordinary circumstances. Ikai and Steinhaus (1961) have demonstrated that significant increases in strength can be elicited following a shout, a shot, drugs, or hypnosis. They argue convincingly that inhibitions are limiting factors in the expression of maximal strength. Remember this point since it will be of great interest when we consider the effects of training on strength. We do not know for certain if we are able to voluntarily call forth all the motor units in a maximal contraction. However, recent evidence suggests that we are unable to do so, since experiments with electrical stimulation have demonstrated greater tension than those with voluntary contractions. It now appears that the maximal potential of the muscle is seldom, if ever, realized in a maximal voluntary contraction. Learned inhibitions from the higher control centers and inhibitory spinal reflex mechanisms seem able to attenuate the expression of force.

Strength training allows the resetting of the inhibitions in such a way as to allow a greater proportion of the maximal force to be available in a voluntary contraction. Ikai and Steinhaus, 1961.) Coaches attempt to get their athletes psyched up so they can temporarily operate beyond normal inhibitory levels. Various drugs have also been used in an attempt to do away with inhibitions. While no one would argue against the use of strength training in athletics, many would question the efficacy of drugs and extreme motivation, both for obvious ethical reasons and because they could upset the normal patterns of skilled performances.

At the cellular level, strength, or the generation of force, must be the result of the tension developed at the cross-bridges within the sarcomeres. Indirect proof of this is found in the relationship between strength and the cross-sectional area. The greater the cross-sectional area (girth) of a muscle, the greater its potential for fibrils, filaments, and cross-bridges. The contractile potential of a muscle could then be described in terms of available actin and myosin filaments.

The relative position of the cross-bridges could help explain why the stretched muscle generates more tension than the partially contracted or shortened muscle. At a grosser level it would also appear that a stretch would remove slack and effectively transfer tension to the tendon. Furthermore, it seems likely that elastic potential energy can be stored in a stretched muscle and then used for the performance of positive work. (Thys, Faraggiana, and Margaria, 1972.) Whatever the case, athletes often place a muscle on stretch immediately before release in events such as the javelin or discus throw. Thus, the increased tension developed at greater than resting lengths may be the result of the removal of slack, elastic energy stored during the stretch, and the favorable alignment of the cross-bridges.

Strength Training

We have already alluded to possible changes in inhibitions due to strength training. However, we should not ignore the possibility of cellular changes. Perhaps the most vivid adjustment to such training could be the development of new muscle fibers, although such a hypothesis is held untenable by most researchers. Van Linge (1962) performed an unusual experiment that could shed some light on the adaptive potential of skeletal muscle. He denervated the triceps surae muscle in rats, and replaced it with a transplanted plantaris muscle which weighed one-fifth that of the triceps surae. The transplantation forced the plantaris to assume a tremendous work load, and heavy training brought about some unusual changes. The trained muscle almost doubled its weight and tripled its force. Furthermore, training stimulated the formation of *new* muscle fibers, strong protein synthesis, and the growth of connective tissue.

While the research literature does not support the development of new muscle fibers as a result of ordinary strength-training procedures, it does support the addition of the contractile proteins actin and myosin. Rats exposed to endurance training experienced the enzymatic changes mentioned in Chapter 3, along with a surprising *decrease* in contractile protein. On the other hand, rats exposed to weight-lifting experiences developed a rise in the concentration of contractile protein and a *fall* in sarcoplasmic (enzyme) protein. (Gordon, 1967.) Thus, the effect of strength training would likely be the addition of contractile protein and not the addition of new fibers. The fact that the specific stimuli of strength training allowed a decrease in the endurance enzymes should not go unnoticed, especially since some coaches and trainers suggest that both strength *and* endurance are developed with strength training. The hypertrophy of a muscle as a result of strength training would be caused by increased protein as well as by the development of connective tissue.

Strength training seems to be the product of tension and, to a certain extent, repetitive movements. Strength can be increased via voluntary isometric or isotonic contractions, or in electrically stimulated involuntary contractions. Changes due to the involuntary contractions only occur when the shortening of the muscle is resisted. However, when compared with the voluntary contractions, the electrical stimulation proved inferior to typical isometric or isotonic methods. These findings support the dual interaction hypothesis mentioned earlier. Voluntary contractions enhance motor pathways and reduce inhibitions while they stimulate the cellular changes. Involuntary (electrical) contractions probably enter via the motor neuron and affect the cellular changes, but they fail to influence the learning processes in the CNS. Thus the effects of voluntary strength training combine cellular and central nervous system (CNS) changes. (Massey et al., 1965.)

Isometric contractions have not provoked the strength gains or hypertrophy that typically result from isotonic programs. However, studies have usually compared a few brief maximal isometric contractions with a full program of isotonic exercise. It is difficult to equate the two forms of effort since isometric contractions involve a force but no distance, hence no measurable work. To date, no researcher has equated tension-time or energy cost in a fair comparison of the two modalities. Research has shown that lifting a weight that permits but 5–10 contractions (ten RM or repetitions maximum) results in a strength gain. When such a program is repeated two or three times a day, every other day, the results seem to be optimal (Berger, 1962). Six-second isometric contractions exceeding one-third of the maximal strength retard muscular atrophy, while contractions at or exceeding two-thirds of the maximal strength provide an isometric training stimulus (Hettinger and Müller, 1953). More recent evidence suggests that several repetitions (five to ten) of short maximal contractions provide the optimal isometric stimulus. (Müller and Rohmert, 1963.) Later in this monograph we will suggest some cautions in the prescription of heavy resistance or maximal isometric contractions for adults.

Isometric contractions offer maximal resistance, but only at a specific angle, and isotonic contractions offer a range of motion with a diminishing resistance. The features of maximal resistance throughout the range of motion are combined in the *isokinetic* principle of training. Through hydraulic or electronic systems the load is adjusted to provide maximal resistance over a wide range of motion. Coaches and physical educators need not despair at the cost of such devices since all that is needed is an elegant feedback system such as the human nervous system. Pupils can work in pairs to provide resistance for each other. As one performs a maximal forearm flexion (concentric contraction) the other resists the movement and experiences a strong eccentric contraction. A little imagination will suggest many uses for a technique that has been termed *proprioceptive neuromuscular facilitation* by physical therapists. The facilitation results when the resistive stretch activates the muscle spindles, and thereby recruits additional motor units. Thus, the isokinetic principle insures the development of strong training contractions that are forced to remain maximal over a wide range of movement. (Rosentzweig and Hinson, 1972.)

Whatever the training technique or stimulus employed, it is likely that strength training is specific to the movements used as well as to the motor units involved. The training somehow stimulates the synthesis of contractile protein, synthesis that typically occurs in the ribosomes of the sarcotubular system. The training also seems to lead to changes in motor pathways and the reduction of CNS inhibitions, both in the higher

TABLE 4.1. EXERCISE PRESCRIPTIONS FOR MUSCULAR STRENGTH AND ENDURANCE

Strength	6–8 RM[a]	3 sets	every other day
Endurance	Over 10 RM[a]	3 sets	every other day

[a] RM: Repetitions maximum (for strength increase resistance when repetitions exceed RM, and for endurance increase repetitions).

centers and the spinal reflex mechanisms. An increase in connective tissue and a possible decrease in endurance enzymes are also associated with strength training.

ENDURANCE

Endurance Training

We have already discussed the cellular basis of endurance and the effects of such training on aerobic enzymes and the fuels for contraction. We mention it again, only to call attention to the differential and specific effects of such training. Strength training calls forth the synthesis of contractile proteins while endurance training stimulates the development and activity of aerobic enzymes. Gordon (1967) reported a surprising drop in contractile protein with endurance training, and a drop in endurance protein with strength training. It appears then that we should expect to get what we train for, and little else. Furthermore, since endurance or repetitive contractions can be accomplished by a portion of the total motor units, we can expect the training stimulus to influence only those units and fibers involved.

The strength or endurance exercises employed should be designed to engage muscle groups in the movement patterns involved in the sport or activity. Generalized training for strength or endurance seems an ill-advised and inefficient use of time. Training of the bicep will not help the tricep. Training for strength will not develop endurance optimally, and the opposite is also true. Many track and swimming coaches have appreciated the value of training at race-pace. This ensures the overload of the specific muscles and the specific metabolic pathways to be utilized in competition.

Before we continue this discussion it is necessary to evaluate the relationship between strength and endurance. We have already differentiated them by virtue of their cellular bases as well as by the very specific effects of training. However, their relationship is more complicated than that. Nevertheless, it is not necessary to devote much time to the complications since the evidence for specific training changes favors specific training experiences. However, when endurance is measured by using the same load for all subjects (absolute endurance) we find strength and endurance *are* highly related. When endurance is tested using a percentage of each subject's maximal strength (relative endurance) they are not correlated. During a series of maximal grip strength trials a stronger subject may experience a greater decrement in performance than a person with lesser maximal strength (Fig. 4.1). These sometimes confusing observations may be explained in part by the nature of the fiber

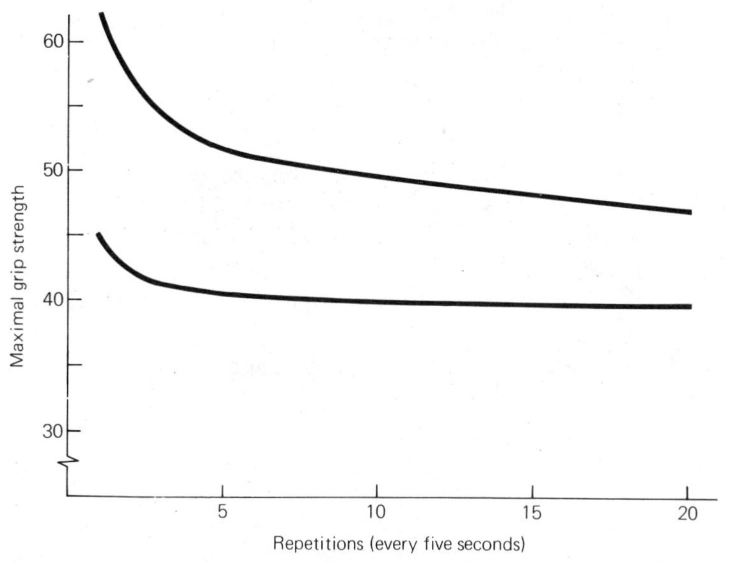

Figure 4.1. Grip fatigue curves. The stronger subject lost a greater proportion of his strength (25 percent) after twenty repetitions. The other subject, a tennis player, was able to maintain a higher percentage of *her* maximal strength. After 20 repetitions she retained more of her strength than did her male counterpart.

types in the muscle. Previous experience (such as years of tournament tennis, for example) may result in well-conditioned muscle fibers that are not particularly strong, but are capable of continuing with a high percentage of the maximal strength. Another consideration is the type of contraction, whether it be isometric or isotonic, and the degree to which it cuts off (occludes) the blood flow to the muscle. An isometric contraction in excess of 60 percent of the maximal voluntary force will occlude the blood flow (Lind, McNicol, and Donald, 1966) and force the fibers to rely on anaerobic energy supplies. The anaerobic properties of the fast-twitch fiber should be helpful in such isometric (static) contractions. Isotonic exercise typically allows the flow of oxygen-carrying blood between contractions. More will be said regarding the relationship of local muscular endurance to human performance in later sections.

Anaerobic Training

We have discussed the possible effects of training on the anaerobic energy sources and pathways (see Chapter 3), but have not yet indicated the type of training to use, nor have we alluded to the interesting interaction between anaerobic training and strength. All evidence points to the need to *overload* a system if such adaptation is to take place. The phosphagen compounds adenosine triphosphate and creatine phosphate (AT and CP) should be severely depleted within 10–15 seconds of maximal effort. Rest periods should allow sufficient recovery to permit maximal overload of those muscle groups to be used in the sport under study. Astrand and Rodahl (1970) suggest that the rest periods between maximal work bouts last for several minutes.

The effect of training on the lactate portion of the anaerobic system is still under investigation. If, in fact, it is subject to such an effect, it would likely result from periods of maximal effort that overload the potential of the system. Maximal work bouts lasting about 1 minute should be followed by 5-minute rest periods. Four or five of these maximal efforts should prove difficult for any athlete. They are so physically and psychologically demanding that they should not be introduced until a strong aerobic foundation has been provided.

The maximal efforts required to tax the anaerobic potential may also serve to enhance muscle strength since the tension developed in the rapid, forceful contractions is sufficient to provide a training stimulus. This may explain the findings noted earlier (see Chapter 2; and Costill et al., 1968) regarding the relationship between anaerobic power and leg strength. On the other hand, it is *quite unlikely* that typical strength-training programs will provide the anaerobic overload necessary to elicit changes in the phosphagen or lactate systems. (Dainty, 1971.)

By now you should realize that specific factors differ in trainability. While it is unusual for one to double his maximal voluntary strength, it is not unusual to see muscular endurance increased many times. We have seen a subject increase his performance in a 25-pound elbow-flexion test from under 50 to over 1000 repetitions. The trainability of the anaerobic power has yet to be fully explored. Thus, it is important for the coach or teacher to determine the specific demands of the sport or activity and to design specific training experiences to meet those demands. Sports demanding the repetition of submaximal contractions should be preceded with endurance-type training. Those that require brief, all-out efforts should include considerable anaerobic training, and those few activities that require high levels of maximal strength should include specifically designed strength-training experiences. All training programs should involve those muscle groups to be used in the activity, in the manner in which they are to be utilized.

Speed of Movement

We should mention speed of movement here because it is one of the most exciting factors in sport and because of its relationship to strength. Before we can elaborate upon their relationship, we should examine our terms and define them more precisely. *Speed of movement* can logically be divided into two components, reaction time and movement time. *Reaction time*—the period from the presentation of a stimulus to the beginning of a movement—can be considered a function of the nervous system. Since we do not seem to be able to accelerate the transmission of the nervous impulses, any improvement in reaction time must be brought about by a greater awareness of appropriate stimuli and the repetition of appropriate responses in order to reduce central nervous system (CNS) processing time (i.e., decision time).

Movement time—that time which elapses from the beginning to the end of the movement—may be improved (decreased) by a program of strength training. Significant improvements in movement time have been associated with gains in strength of the related muscles, and increased projectile velocities have been reported following related overload programs. The strength training can increase the contractile protein. The effects of such training would be most evident in events involving heavy resistances and maximal velocities, such as the shot put or hammer throw. However, even the velocity of the baseball throw has been improved by training with weighted balls and pulley weights. (Brose and Hanson, 1967.)

In an isolated muscle, the velocity of shortening is greatest when resistance is absent. As the resistance is increased the velocity of shortening decreases. (Hill, 1964.) Though training would not be likely to influence the velocity of the unloaded movement, it does seem to allow a higher velocity at the pretraining load or a higher load at the pretraining velocity. An obvious example of the latter is the weight lifter who is able to move (lift) a greater load after training. Thus, it appears possible to modify the slope of the force-velocity relationship (Fig. 4.2).

We should not conclude from the above that continued increases in strength will lead to continued improvement in movement time. Nor should we ignore possible movement-time limitations such as a predominance of slow-twitch fiber types and limits in intrinsic speed and neuromuscular coordination. The extent to which strength training is capable of improving movement time has yet to be fully explored. It is safe to conclude, however, that speed—like strength or endurance—is extremely task specific. The speed of arm movement is not necessarily related to the speed of leg movement. Some may be quick with their hands while others may be capable of rapid leg movements. Skill and strength training both seem capable of influencing the time required to complete a given task.

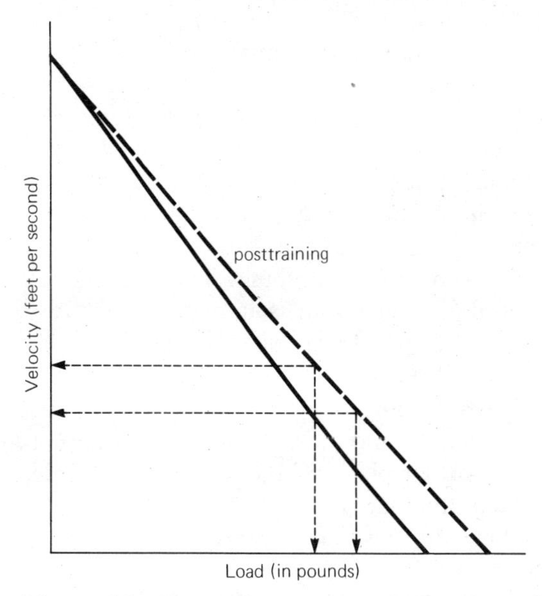

Figure 4.2. Strength training and the force-velocity relationship. It is likely that the benefits of strength training would be more pronounced in events involving loaded or resisted movements.

MUSCULAR FATIGUE

Fatigue is a complex phenomenon that has, to date, defied explanation. It is essentially a subjective sensation which is related to complex physiological as well as psychological factors. While some have argued that the term *fatigue* should be reserved for the subjective sensation alone, Simonson (1971) proposes that the term be used for all the processes that result in a performance decrement. When speaking of muscular fatigue we will focus our consideration on the physiological events that limit muscular contractions. This should not suggest, however, that psychological factors are of minor consequence in human muscular performance. In fact, it may be that most of us seldom, if ever, reach the physiological limitations we shall discuss. These limits may only be attained by a few dedicated athletes and others who are thrust into extremely demanding conditions. On the other hand, the points we shall mention may operate to reduce the intensity or quality of performance.

Muscular fatigue may occur at one or several points. One site of fatigue often mentioned in earlier sources was the myoneural junction or motor endplate. We have all seen the experiment employing the nerve-muscle preparation from the frog, where stimulation of the nerve eventually fails to elicit muscular contractions. Movement of the electrodes to the muscle results in contractions, presumably proving that the myoneural junction is the site of fatigue. However, several lines of reasoning cast doubt on that hypothesis. The nerve-muscle preparation is removed from its normal sources of supply. Also, the rate or magnitude of stimulation may be above that encountered within the physiological system. Finally, Merton (1954) presented evidence indicating that the fatigued muscle was receiving stimulation, but was incapable of exerting tension. Using surface electromyography (EMG), Merton recorded muscle action potentials from fatigued abductor pollicus muscles. Based on these findings, the myoneural junction seems an unlikely site of fatigue.

Another possible site or cause of fatigue is the ATP available to supply the energy for contractions. However, ATP does not seem to be fully depleted during fatigue, perhaps because it is an important energy source for other cellular functions.

If ATP supplies are low or unavailable to the contractal units of the muscle, actin and myosin join in low-energy complexes called *rigor complexes* (Murray and Weber, 1974). Although it is far too early to relate this laboratory finding to human muscular fatigue, it is appealing to speculate about the relationship of these rigor complexes to the outward manifestations of fatigue in events such as wrestling and the 440-yard dash.

CP, does seem to be more severely depleted during muscular fatigue. As you know, CP is either a secondary source of high-energy phosphate bonds or a source of energy for the regeneration of ATP (CP + ADP → C + ATP). Rapid, repetitive contractions deplete CP faster than support systems can provide energy for ATP resynthesis. However, Karlsson and Saltin (1970) reported results from exhaustive bicycle ergometer studies that exonerate phosphagen stores as the likely cause of muscular fatigue. CP was depleted to the same level after 2–3 minutes, whether the exercise was exhaustive or not, and, when the work was continued, there was no tendency for a further depletion.

These findings suggest a shift of attention from phosphagen to other substrates or by-products of cellular metabolism. We have already noted that glycogen is the factor that determines long-term performance capacity at high submaximal loads. In brief exhaustive efforts, the ability to accumulate lactate may be the cause of muscular fatigue. (Karlsson and Saltin, 1970.) The acidic effects of lactate accumulation could overwhelm the buffering capacity of the cell and lead to changes in the *p*H that could alter enzyme activity. While no conclusive evidence exists to support such a possibility, Osnes and Hermansen (1972) have reported a decrease in muscle *p*H from 6.93 at rest to 6.40 following maximal exercise of short duration.

Yet another possible site of fatigue is the poorly understood link between excitation and contraction. This link involves the passage of the impulse in the sacrotubular system and the activation of ATPase activity. Rapid contractions could deplete calcium and eventually block the splitting of ATP and the formation of active actin-myosin linkages. (Sandow, Taylor, and Preisier, 1965.) It is interesting to note that although the calcium-uptake capacity of sarcotubular vesicles is less in red than in white muscle fibers, endurance training had no effect on the calcium capacity of guinea pig muscle. (Barnard, Edgerton, and Peter, 1970.) Thus, the excitation–contraction link remains a poorly understood, but potentially crucial, factor in the complicated study of muscular fatigue.

The Two-Factor Theory

Our discussion to this point has not adequately considered the intensity and duration of the contractions that lead to impaired contractile ability. Missiuro, Kirschner, and Kozlowski (1962) studied the electromyographic manifestations of fatigue and concluded that two mechanisms were involved. Peripheral fatigue of the contractile apparatus was produced by intense exertion of short duration while CNS fatigue resulted from long-duration, low-intensity contractions. We have already mentioned several possible sites of peripheral fatigue (ATP, CP, *p*H, glycogen, calcium). CNS fatigue may be the result of low blood glucose, a limited oxygen supply, or both. Whatever the cause it seems that the fatigue that results from low-intensity work of long duration is due to a decrease in discharges from the CNS. But in intense effort of short duration the CNS *increases* the stimuli to the muscle in an apparent effort to recruit additional motor units. These findings support the glucose-sparing theory mentioned earlier. During prolonged effort, muscular metabolism of glucose is avoided in order to reserve an adequate supply for nervous tissue.

In his exhaustive treatise on fatigue, Simonson (1971) suggests the involvement of four basic processes.

1. accumulation of substances (e.g., lactate)
2. depletion of substances necessary for activity (ATP, CP, glycogen)
3. changes of the physiochemical state of the substrate (*p*H)
4. disturbances of regulation and coordination (blood glucose, CNS)

The reader is urged to consult this excellent source for a thorough analysis of the physiology of fatigue.

Finally, it is important to reemphasize that most individuals probably experience fatigue before they reach the limits mentioned. The discomfort or pain of repetitive contractions, along with learned inhibitions, may result in the termination of effort before physiological limits are reached. Repetitive training experiences may allow athletes to approach these limits either by the resetting of inhibitory levels, the blocking of painful stimuli from conscious awareness (habituation), or through the interaction of complex physiological *and* psychological changes resulting from training.

PART II PHYSICAL ACTIVITY: SUPPORT SYSTEMS

Having dealt with the control of movement, the contractile process, and the sources of energy for contraction, let us now consider the important systems that serve to supply and augment the working muscles. In the chapters which follow, we shall consider respiration and gas transport, the blood and circulation, and the influence of the hormones of the endocrine system. Also included in Part II is a review of important environmental factors such as heat, cold, altitude, pressure, and air quality which influence and interact with the support systems.

In Chapters 5–8, we shall discover the answers to the following questions: Do you really run out of wind, or is your lung capacity an important contributor to performance capacity? What are the effects of training on the cardiorespiratory systems, and what portions of these support systems limit our capacity for maximal performance? How does exercise in the heat affect our circulation, our work capacity? How does altitude alter the dynamics of respiration? Does physical fitness improve our ability to tolerate heat or high altitude? What are the effects of exercise on the endocrine glands and what are the effects of various hormones on performance?

To answer these and other questions, it is necessary for us to study systems that support and supply the working muscles and allow continued physical activity.

CHAPTER 5 RESPIRATION AND GAS TRANSPORT

We have all experienced the distress associated with the "I ran out of wind" feeling. Just how important is our ability to supply air to the lungs in physical activity? Do respiratory phenomena limit our capacity for physical activity?

LUNG VOLUMES

Recently, one of my undergraduate students decided to compare athletes, vocalists, and horn players to see which group possessed the largest *vital capacity* (VC), that is, the greatest ability to bring air into the lungs. It may come as a surprise, but the musicians did rather well, many having larger lung capacities than did the athletes. In fact, one outstanding athlete (a distance runner) had a VC well below that predicted for his height and weight. Such static tests of lung capacity are not very useful for the study of the functional ability of the normal respiratory apparatus. VC increases with body size and age until the third decade (Fig. 5.1), then decreases as the residual volume (RV) increases. The RV represents that portion of total lung capacity (TLC) that cannot be exchanged or exhaled. Training—especially during adolescence—seems to improve both VC and TLC because of its effect on the dynamics of pulmonary function.

Dynamic tests of pulmonary function provide greater insight into the functional capabilities of the respiratory apparatus than do static tests. An asthmatic patient may retain his VC, but when he attempts to move that volume of air quickly—as he must during exercise—he will be severely limited. In the determination of the *forced expiratory volume,* the subject must inspire maximally and then exhale as quickly and completely as possible. Untrained respiratory muscles and airway constriction (e.g., asthma) limit performance in dynamic tests as well as performance in sport. The *maximal voluntary ventilation*

		♂	♀
vital capacity ♂ = 4.8 liters ♀ = 3.2 liters	inspiratory reserve volume	3.3	2.0
	tidal volume	0.5	0.5
	expiratory reserve volume	1.0	0.7
residual volume (in liters)		1.2	1.1
	total lung capacity	6.0	4.3

Figure 5.1. Lung volumes. Although residual volume increases with age, active adults seem to have a smaller residual volume (RV) than their inactive counterparts.

test (MVV test) is another useful dynamic test for measuring the overall capacity of the breathing apparatus to pump air. The subject breathes as rapidly and deeply as possible for 15 seconds, and the volume of expired air is collected and measured (men average 140 liters per minute, women about 100 liters). Factors that may limit the MVV include changes in the elasticity of the lung because of age or disease, untrained respiratory musculature, and airway resistance caused by contraction of the smooth muscles of the bronchial tubes. Bronchoconstriction—so characteristic of the asthmatic—may also result from the irritation of cigarette smoke or other airborne pollutants. The insidious effects of air pollution will be discussed later (see Chapter 7), as will other factors that affect human performance (see Chapter 10).

Pulmonary Ventilation

The pulmonary ventilation ($\dot{V}$), or the amount of air inspired per minute, ranges from 6–8 liters per minute at rest to 120–180 liters or more during maximal exercise ($\dot{V}_{max}$). Mechanically, this movement of air is caused by the activity of the diaphragm and external intercostal muscles at rest. In exercise, the scaleni and sternocleidomastoids assist the aforementioned muscles during inspiration, while the internal intercostal and abdominal muscles assist with the forceful expiratory effort. As the demands of exercise call forth an increase in $\dot{V}$, the respiratory muscles demand an ever-greater share of the oxygen brought into the lungs. The oxygen cost of ventilation may increase from 1 percent of the total oxygen uptake at rest to 10 percent during maximal effort. The work of the muscles increases to overcome elastic and airflow resistance that result from deeper and more rapid inhalations and the need for rapid forceful exhalations. Physiologists have sought to determine the point of diminishing return, that point at which increased pulmonary ventilation no longer contributes additional oxygen for use by the exercising muscle, where the extra oxygen supplied is needed to drive the breathing muscles. However, a reported $\dot{V}_{max}$ of 200 liters (Saltin and Astrand, 1967) casts doubt on earlier reports that suggested a functional limit of 120 liters. Such suggestions may have been based on studies utilizing collection devices (mouthpieces and tubing) which added to the airway resistance and the oxygen cost of breathing. Breathing devices that increase the resistance to the passage of air severely limit physical working capacity and increase the oxygen cost of respiration. (Thompson and Sharkey, 1966.)

It has long been known that a high ventilatory capacity is required to achieve a high aerobic capacity (maximal oxygen intake). Consequently, many investigators have concluded that $\dot{V}max$ was a limiting factor in maximal performance. However, since oxygen tension in the arterial blood (pO_2) is only slightly reduced and the respiratory muscles are not maximally taxed during maximal work loads, we can conclude that pulmonary ventilation is probably *not* a limiting factor for the normal individual at sea level or at low altitudes. Patients with asthma or emphysema are certainly limited, as are normal individuals at moderate or high altitudes, where the reduced partial pressure of oxygen calls forth a greater need for ventilation (see Chapter 7).

Control of Ventilation

The rate of respiration (f) and the volume per breath or tidal volume (TV) are continuously controlled to provide the necessary pulmonary ventilation.

$$\dot{V} = f \times TV$$

Typically, we leave this control to the autonomic nervous system. However, in singing, speaking, and in some forms of sport we consciously control the pattern of respiration. The singer must alter the rhythm and frequency of breathing to fit the artistic demands of the music. The downhill skier must consciously exhale to counter the breath holding that results from excitement and static racing positions. The weight lifter takes a deep breath prior to making an expiratory effort against a closed glottis (Valsalva maneuver) during a maximal lift. But the conscious control of ventilation is limited, as each of us knows from having attempted to hold our breath for long periods.

The control of ventilation is too important to be left to the fickle nature of our consciousness. Instead, chemical and nervous mechanisms join in an effort to adjust air intake to cellular demands for oxygen. Chemical control mechanisms involve the chemoreceptor sensing of carbon dioxide, hydrogen ions, and arterial oxygen saturation. The production of carbon dioxide is directly related to muscular metabolism. Increases in hydrogen ion concentration and decreases in the *pH* of the blood are only noted during strenuous exercise. Arterial oxygen saturation only declines noticeably during near-maximal exercise, and then the decline may depend on the exercise capabilities of the subject (a 2-percent decline for untrained and 10 percent for trained athletes; Rowell et al., 1964).

Evidence for nervous control of $\dot{V}$ arises from observations of the rapid increase in pulmonary ventilation at the onset of exercise. Since these adjustments occur before changes in carbon dioxide could reach the chemoreceptors, it is believed that they are transmitted via the nervous system to respiratory control centers in the medulla. While a portion of the nervous increase in respiration may be explained as a conditioned reflex involving anticipation and excitement prior to athletic competition or a treadmill test, impulses arising from receptors located about the joints are also involved. Thus, the immediate increase in $\dot{V}$ may be caused by nervous activation of the inspiratory center, and the further increase during exercise may involve both chemical and nervous control. It can easily be demonstrated that $\dot{V}$ increases during exercise in relation to metabolic activity, most specifically to the carbon dioxide production. Furthermore, forced breathing or hyperventilation at rest will markedly increase breath-holding time. The hyperventilation seems to blow off carbon dioxide, and thereby reduces the respiratory stimulus. Therefore, carbon dioxide can be seen as the major factor in the chemical control of ventilation (Fig. 5.2).

The respiratory mechanism seems to become more efficient following endurance training. The athlete does less breathing work to maintain his oxygen supply at a given level; that is, he breathes in less air to extract the oxygen required at a given work load. While this efficiency may be due in part to the effects of training on the respiratory musculature as well as oxygen transport and utilization, it may also be related to

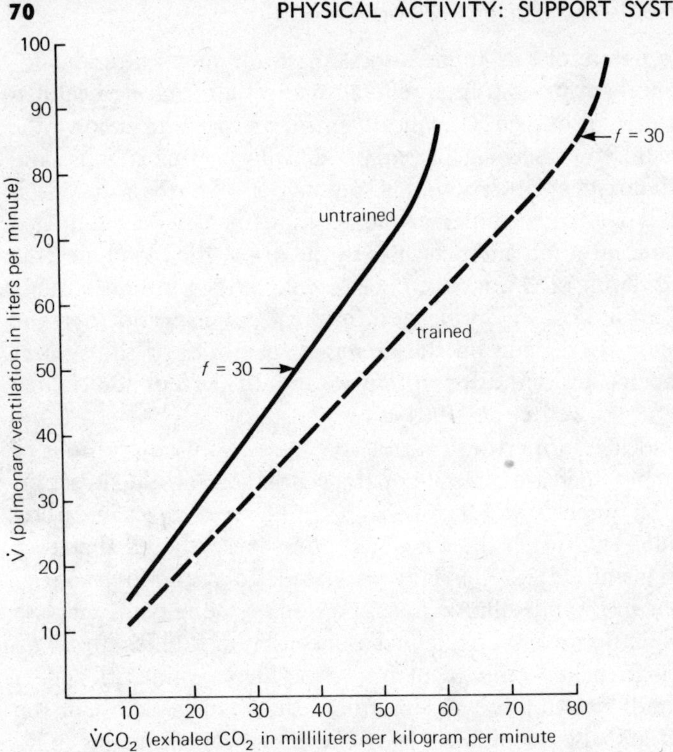

Figure 5.2. The relationship of pulmonary ventilation to carbon dioxide production. In addition to showing the effect of CO_2 on ventilation, this graph illustrates how a trained individual is able to move more air at a given rate of respiration ($f = 30$ breaths per minute). Pulmonary ventilation is the product of respiratory rate and the tidal volume ($\dot{V} = f \times TV$). Ventilation is also related to exercise intensity (oxygen intake) in a linear fashion. Thus, ventilation is adjusted to suit oxygen needs during exercise.

neurogenic changes that reduce the neural component of the respiratory drive. Hence, a diminished respiratory rate, for example, could allow a longer diffusion period in the lungs and a greater extraction of oxygen per liter of inspired air. The repetitive stimuli of training could result in the relearning or resetting of control mechanisms or the reduction of such stimuli to higher control centers. These fascinating aspects of the training-learning interaction have yet to be fully explored. However, they cannot be ignored as possible psychological components of the training effect whenever neurogenic controls are involved, as in respiratory and cardiovascular control.

Diffusion

The respiratory system functions to transport oxygen to the blood and carry away carbon dioxide. The gaseous exchange takes place across the alveolar membrane and through the capillary wall. The physical process of diffusion—where molecules move from an area of higher concentration to a lower one—explains the movement of oxygen into and carbon dioxide from the blood. Oxygen has a higher molecular activity or partial pressure in the atmosphere. As you might expect, the partial pressure of a gas depends on the percentage of that gas in the total sample (20.93 percent $\times$ 760 mm Hg = 159 mm Hg partial pressure or pO_2). Thus, the pO_2 is lower in the lungs due to the oxygen mixing with air in the airway dead spaces (which are not involved in gas exchange) and the addition of water vapor to the gas mixture. By the time it reaches the alveoli—the air sacs where the exchange takes place—the pO_2 is down to about 100 mm Hg. The movement into the capillary is facilitated by the fact that the pO_2 of mixed venous blood is down to 40 mm. The diffusion into the blood takes about half the time it takes for blood to pass through the pulmonary capillary at rest (1 second). Since the blood may flow through the capillary at twice that rate during maximal effort, there is still time for the pO_2 to reach equilibrium with the arterialized blood (pO_2 = 100 mm Hg). The oxygenated blood then travels to the tissue capillaries where the oxygen again passes to a region of lower pressure. The oxygen needs of the tissues depend upon their activity. During vigorous exercise the venous blood draining the active muscles will have a lower oxygen pressure (pO_2) than it would at rest.

Carbon dioxide diffuses far more rapidly than does oxygen. The pCO_2 is highest in the venous blood draining active tissues, lower in the expired air, and lowest in the atmosphere (pCO_2 = 0.2). It enters the tissue capillary as oxygen is leaving, travels to the pulmonary capillary where it leaves as oxygen is entering, and is finally exhaled.

Several factors affect this diffusion capacity ($\dot{D}_L$) and the ability to supply oxygen during vigorous physical activity. The $\dot{D}_L$ depends upon the number of pulmonary capillaries that are open and supplying blood. It depends also on the hemoglobin concentration and the red blood cells in the blood perfusing those capillaries. On the other side of the membrane, the diffusing capacity depends on the functional surface area available. Both alveolar ventilation and capillary perfusion may be increased during exercise but the resulting increase in $\dot{D}_L$ seems to reach a plateau at light work loads and increase only slightly thereafter. (Holmgren and Astrand, 1966.) Disease states can affect $\dot{D}_L$ through a reduction of surface area (e.g., alveolar breakdown in emphysema) or a thickening of the alveolar membrane (e.g., pulmonary fibrosis).

Thus, the diffusing capacity can be viewed as an indicator of the dimensions and efficiency of gaseous exchange. Values range from 25 ml of oxygen (per mm Hg pressure gradient per minute) at rest to 80 ml of oxygen during maximal effort. (Margaria and Cerretelli, 1968.) Generally, the maximal $\dot{D}_L$ is greater in those with a high aerobic capacity, but it does not seem to be a limiting factor in the maximal oxygen intake (except above 5000 feet; Johnson, 1967) since the pO_2 of the arterial blood remains high during vigorous effort. The effects of age and inactivity on maximal $\dot{D}_L$ are not entirely clear, but Donevan et al. (1959) have reported a decline with age that may be caused by a reduction in the size of the vascular bed (perfusion) or to some change in the alveolocapillary membrane. (Cherniack and Cherniack, 1962.) The effects of physical training on $\dot{D}_L$ are not well established despite the attention given the subject in recent years. (Maksud et al., 1971.) Competitive swimmers do seem to enjoy a high $\dot{D}_L$ which may result, at least in part, from restrictions placed on respiration while swimming.

GAS TRANSPORT

We are aware that oxygen can be dissolved in a fluid and utilized to support aquatic life. But if we were to depend on the small amount of dissolved oxygen which can be carried in the blood plasma (0.29 ml per 100 ml of blood), we would be in serious trouble. Hemoglobin increases the oxygen-carrying capacity of the blood about seventy times. Let us examine the dynamics of oxygen transport as a basis for some observations on the relationship of gas transport to exercise performance capacity.

Oxygen Transport

The hemoglobin molecule is a protein composed of four subunits, each containing an iron-bearing *heme* portion. Each of the iron atoms (Hb) can temporarily bind a molecule of oxygen (O_2) in a loose association called *oxygenation.*

$$Hb_4 + 4(O_2) = Hb_4O_8$$

Under normal atmospheric conditions, when the alveolar pO_2 is about 100 mm Hg, the arterial blood contains about 19.8 ml of oxygen per 100 ml of blood. The hemoglobin carries about 19.5 ml and the remaining 0.29 ml is dissolved in the plasma. While men average 16 gm of hemoglobin per 100 ml of blood, women have but 14 gm. Each gram of hemoglobin can carry up to 1.34 ml of oxygen per gram, so the blood is capable of carrying over 20 ml of oxygen per 100 ml. However, this can be accomplished only when the alveolar pO_2 is above 158 mm Hg; for example, when breathing oxygen.

The oxygen content of venous blood drops from 19.8 ml per 100 ml (arterial) to about 15.2 ml at rest. This is an arteriovenous difference of 19.8 — 15.2 = 4.6 ml per 100 ml, the amount of oxygen removed from the blood for use by the tissues. At this point, the blood has lost about 25 percent of its oxygen load, about 4.6 ml per 100 ml have been removed, and the total saturation has been reduced. The degree to which the oxygen is replaced during passage through the pulmonary capillaries, to which the oxygen saturation of hemoglobin returns, is dependent on several factors.

The prime factor responsible for the saturation of hemoglobin with oxygen is the partial pressure of oxygen. Figure 5.3 illustrates the percentage of oxygen (O_2) saturation of hemoglobin (Hb) in relation to pO_2. The right ordinate also illustrates the amount of oxygen combined with hemoglobin at different levels of saturation. The characteristic shape of the saturation curve

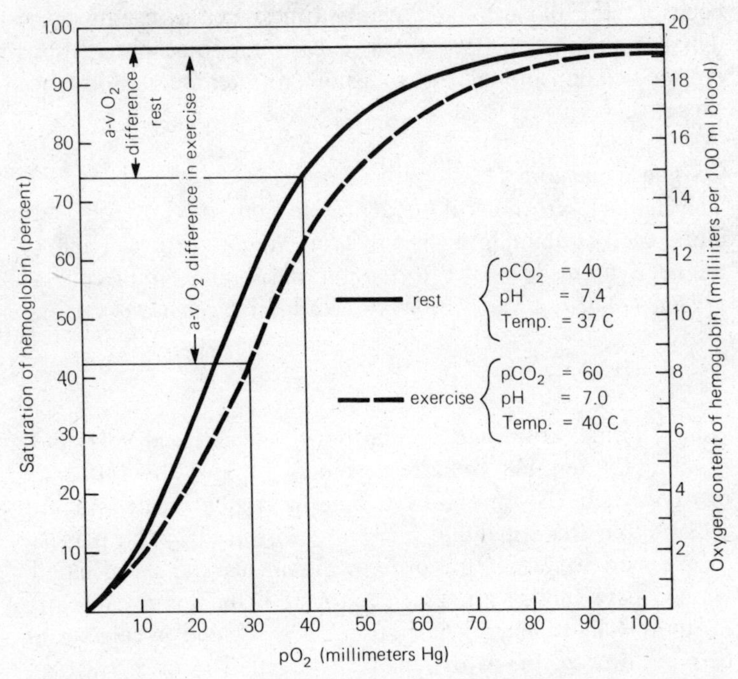

Figure 5.3. The oxygen hemoglobin dissociation curve. Changes in pCO_2, pH, and temperature during exercise cause the curve to shift to the right and assist in the delivery of oxygen to the exercising muscles.

tells us a great deal about oxygen transport. An obvious point is that saturation increases as pO_2 increases, and that the saturation tends to plateau at high levels of pO_2. Thus, there is little change in saturation when the pO_2 drops from 100 to 80 mm Hg. This affords the oxygen transport system a wide range of adaptability and performance, even at moderate altitudes where the lower atmospheric pO_2 will reduce the pO_2 in the alveolar air. The steep portion of the curve allows the active tissues to extract oxygen readily from the blood. The hemoglobin in arterialized blood with a pO_2 of 100 mm Hg is about 97 percent saturated and, therefore, will carry about 19.5 ml of oxygen per 100 ml of blood. The pO_2 of the tissues at rest averages about 40 mm Hg. The saturation falls to about 75 percent, and the tissues receive about 4.5 ml of oxygen per 100 ml of blood. During maximal exercise, the muscle pO_2 may drop to 10 mm Hg or below, the saturation to 10 percent, and the arteriovenous difference may climb theoretically to 17 ml per 100 ml (19.5 — 2.5).

However, several other factors affect the oxygen saturation of hemoglobin, and they all are altered during exercise. In Fig. 5.3, our original curve (*dark line*) represents the conditions encountered during rest (temperature, 37 C.; pH, 7.4; pCO_2, 40 mm Hg). During exercise the temperature and pCO_2 rise, the pH falls, and all three shift the saturation curve to the right (*dotted line*); that is, they all operate to alter the saturation at a given pO_2. Because of the plateau at the upper end of the curve this shift does not have too profound an effect on saturation, although the effect is detectable during maximal effort. Rowell et al. (1964) measured the effect of maximal exercise on the saturation of arterial blood. Saturation dropped only 2.4 percent in sedentary subjects, 4.4 percent after training, and over 10 percent in trained athletes. So it is clear that the shift of the saturation curve could significantly alter oxygen transport during maximal effort.

During maximal effort the temperature may rise to 40 C. The pH may drop as low as 6.8 (Osnes and Hermansen, 1972) due to the production of lactic acid (H^+ ions) as well as carbon dioxide ($CO_2 + H_2O \rightleftharpoons H_2CO_3 \rightleftharpoons H^+ + HCO_3^-$). A favorable aspect of the shift may also be seen. It makes more oxygen available to the tissues at a given pO_2 since any shift will result in greater changes on the steep portion of the curve. Thus, the shifting of the saturation curve may appear to be a disadvantage in that it can lead to a reduction in saturation for trained athletes at maximal effort, but its overall effect is an improvement in oxygen delivery to the tissues.

Carbon dioxide transport takes place principally within the red blood cells. It combines with water, and, with the help of the enzyme carbonic anhydrase, forms H_2CO_3, which then dissociates to form hydrogen ions and a bicarbonate ion (HCO_3^-). Additional carbon dioxide is also carried in association with the protein portion of the hemoglobin molecule. This competition for space on the hemoglobin molecule facilitates the dissociation of oxygen in the tissue capillary as well as the unloading of carbon dioxide in the pulmonary capillary. So we see gas transport as an elegant system that seems to hasten the discharge of oxygen at the tissue level and the removal of carbon dioxide in the lungs.

PHYSICAL ACTIVITY AND THE ACID-BASE BALANCE

You are aware that physical activity increases carbon dioxide and lactic acid production, and that both of these by-products lead to the formation of hydrogen ions. In an ordinary solution, the addition of hydrogen ions would rapidly lower the *p*H as the solution becomes more acidic. However, the blood is no ordinary fluid, and the exercising organism seems able to tolerate sizable increases in hydrogen ion concentration without a noticeable change in *p*H. But during vigorous exercise the *p*H of the blood may be forced from the normal 7.4 down to or below 7.0 for brief periods. How does the body attempt to resist this decline, and what can be done to improve our ability to tolerate its effects?

Buffers

We seem to have three lines of defense against marked changes in the acid-base balance: buffers, respiration, and the kidneys. The first two function during exercise and seem to complement each other. The kidneys receive a diminished blood flow during vigorous exercise so they must do their job after the exercise has been terminated.

Buffers accept hydrogen ions and thereby maintain the *p*H at a tolerable level. Let us briefly consider three important buffers that serve to maintain the blood *p*H at or near the 7.4 level. *Proteins* in the blood, particularly the plasma proteins, act as hydrogen acceptors since they ionize in the blood to form negatively charged particles. Thus, the negatively charged portion is able to accept the positively charged hydrogen ion and resist changes in the *p*H. For example,

$$H^+ + \text{protein} \rightleftharpoons \text{H protein}$$

Hemoglobin provides a buffering capacity six times as effective as the protein system mentioned above. Moreover, when oxygen leaves the hemoglobin molecule the molecule becomes a weaker acid and consequently is more effective as a buffer. As with the protein buffers, the addition of an acid causes the two reactions to move to the right.

$$H^+ + Hb^- \rightleftharpoons HHb$$

These systems continue to function so long as there are weakly dissociated and negatively charged protein or hemoglobin molecules available to accept hydrogen ions.

The third important buffer system involves *carbonic acid and bicarbonate.* This system can be illustrated in simplified form similar to that used for the protein and hemoglobin systems.

$$H^+ + HCO_3^- \rightleftharpoons H_2CO_3$$

However, we can also illustrate the fact that buffers involve a weak acid and a salt of that acid. The addition of hydrogen ions will cause equation (1) to move to the right and equation (2) to the left.

$$H^+ + HCO_3^- \rightleftharpoons H_2CO_3 \qquad (1)$$
$$Na^+ + HCO_3^- \rightleftharpoons NaHCO_3 \qquad (2)$$

Thus, $NaHCO_3$ stands ready to provide the HCO_3^- ions needed to tie up hydrogen ions and prevent undesirable changes in the *p*H. This buffer system is not particularly effective at the *p*H of blood. But since the H_2CO_3 level depends on the concentration of carbon dioxide, and since the carbon dioxide seems to drive the respiratory apparatus, the carbonic acid-bicarbonate buffer system becomes extremely effective.

Respiration

While the buffers are immediately available for the control of hydrogen ions, respiratory assistance may not begin for a minute or longer. The building levels of carbon dioxide and carbonic acid lead to an increase in pulmonary ventilation and the blowing off of carbon dioxide.

$$H_2CO_3 \rightleftharpoons H_2O + CO_2 \quad \text{(exhaled)}$$

The fact that pulmonary ventilation increases with exercise intensity is well established. The relationship of respiration to the carbonic acid buffer system is a fascinating example of the elegant and closely linked control mechanisms that combine to permit vigorous physical activity. Another example is that of the sudden change in the hemoglobin buffering capacity upon the removal of oxygen, a change that takes place just when a hydrogen acceptor is needed.

The Kidneys

Although the kidneys may be unable to do their part for acid-base balance during vigorous exercise, their role in the regulation of *p*H cannot be ignored. The kidneys serve to remove hydrogen ions finally from the circulation. As a result, urine can often be quite acidic, even painful to void. In the process, the kidneys are careful to reabsorb sodium ions (Na^+) in order to restore the buffer system to normal. While the buffer system may function to temporarily resist changes in the *p*H, the kidneys provide the final solution, that is, elimination.

Summary

Carbon dioxide is directly related to exercise intensity. At light or moderate work loads the buffer systems and respiration are quite able to maintain a tolerable *p*H. However, as increasing exercise intensity demands more anaerobic metabolism and lactic acid formation, the combined carbon dioxide and lactic acid by-products begin to tax the overall buffering capacity. Although experience has shown that training allows one to tolerate a greater lactatae accumulation as well as a lower *p*H, no method has yet been devised to improve the buffering capacity during vigorous exercise.

TRAINING

We will resist the urge to delve into several topics closely related to the material covered in this chapter, such as the effects of altitude on performance, the effects of oxygen inhalation on performance, and the effects of air pollution on athletes. These and other factors related to human performance will be discussed in later chapters. Instead, we will conclude this chapter with a brief account of possible training effects related to respiration and gas transport.

It seems likely that training will influence both the static and dynamic lung volumes as well as maximal pulmonary ventilation. All of these factors are highly correlated ($r > 0.8$; Holmgren, 1967) to overall oxygen transport as indicated by the maximal oxygen intake (aerobic capacity). Remember, however, that these volumes are not believed to be limiting factors in maximal effort for normal, healthy individuals performing at low altitudes. The diffusing capacity at maximal work loads may be improved by training, although the evidence for this is not overwhelming. Diffusing capacity is also highly correlated to the aerobic capacity. Both blood volume and total hemoglobin are increased because of training and *total* hemoglobin is highly predictive of the aerobic capacity ($r = 0.919$; Holmgren, 1967). Neither buffering capacity (alkali reserve) nor plasma protein concentration is affected by training. However, training does seem to allow an increase in the ability to tolerate lactic acid. We will say more about the effects of training in Chapter 6 when we consider the roles of respiration and circulation in cardiorespiratory endurance.

CHAPTER 6 CARDIOVASCULAR DYNAMICS

It is the day before the conference meet. The distance runners go to the college infirmary to receive a reinfusion of their own blood, taken from them four weeks before. The following day they go out to win *all* the distance events and *several* of them even beat the school record for the mile run. . . .

Sounds a bit farfetched doesn't it, a bit too much even for today's competitive society? Of course it is, but it is based on a recent experiment performed by Ekblom, Goldbarg, and Gullbring in Sweden (1973). The experiment was conducted not to gain an advantage in an athletic event but to shed further light on the importance of hemoglobin to the oxygen transport system. The subjects who participated in the actual experiment did not compete in a track meet, but they were able to immediately increase their endurance performance 23 percent and their aerobic capacity some 9 percent. This experiment dramatically illustrates the importance of blood and circulation to the physical working capacity. Let us now consider the blood, the heart that pumps it, and the circulatory system that delivers it, and then conclude with some of the effects of training on the cardiorespiratory system.

THE BLOOD

The blood serves to *transport* foodstuffs, gases, waste products, hormones, antibodies, and heat. It also acts to regulate the acid-base balance. We have already dealt with some of these functions. Let us take a brief look at the cellular and plasma portions of this complex fluid and consider its importance in exercise and training. The cellular components of the blood, including red cells, white cells, and platelets, comprise about 45 percent of the total blood volume. Blood volume averages about 5 liters and constitutes about 7–8 percent of the body weight for a man weighing 70 kg (154 pounds).

White cells, numbering 4,000–11,000 per cubic millimeter, include several types involved in phagocytosis and antibody reactions. The red cells, numbering about 5 million per cubic millimeter, are formed in the bone marrow and survive in the circulation for about 120 days. As we shall see in a later section, red cell production (erythropoiesis) can be stimulated by the hypoxia encountered at altitude. The red cells contain all the hemoglobin found in the blood (about 15 gm per 100 ml). The hemoglobin is lost when the cell completes its life cycle and it must be replaced. However, the iron portion of the hemoglobin molecule is not lost and it can be reused.

Some forms of hemoglobin contain an inherited abnormality. In *sickle cell anemia*, a single misplaced amino acid in a chain of 300 causes the hemoglobin molecule to be very insoluble at a low partial pressure of oxygen. The resulting sickle-shaped red cells hemolyze and cause severe sickle cell anemia. This inherited trait is found only in those with Negroid blood and it usually appears after some event brings about a low oxygen tension (pO_2). For this reason, physiologists have been concerned about the effect of exercise at altitude on black athletes. Prior to the 1968 Olympics in Mexico City, there was considerable concern regarding the effect of the reduced pO_2 on athletes with the heterozygous sickle cell trait. However, the combination of low pO_2 and severe exertion failed to provoke anemic crises despite the large contingent of successful black athletes in attendance.

Platelets, numbering about 300,000 per cubic millimeter are important in cases of injury and clotting. Due to the substances contained within their walls, including epinephrine, norepinephrine, serotonin, histamine, adenosine triphosphate (ATP), and ribonucleic acid (RNA), they are involved in clot formation and retraction. Also, the serotonin seems to cause local vasoconstriction when a blood vessel is injured.

Blood plasma is remarkable, both for the number of things it carries and for the many things it can do. We will deal with the processes of heat transfer and temperature regulation in Chapter 7. Those properties are made possible by the remarkable nature of the unusual fluid that forms the base of blood plasma—water. Water is also known as the universal solvent, but not all the constituents of plasma are in true solution. Some very large protein molecules are suspended in colloidal solution. These proteins are important because they constitute one-sixth of the buffering capacity of the blood and also because they exert an osmotic force that tends to pull water into the blood. Albumin is the plasma protein most responsible for the colloidal osmotic pressure of the blood. The globulin protein fraction is associated with antibody formation. Fibrinogen, the largest of the plasma proteins, is an essential element in the clotting mechanism.

Exercise and Blood Clotting

Under normal circumstances the tendency of blood to clot is inhibited by a series of events that operate to prevent intravascular clotting. This *fibrinolytic* system serves to remove the clots that form when soluble fibrinogen is converted into insoluble fibrin threads. The involved details of the clotting mechanism are beyond the scope of this monograph; however, because of the implications for those of us involved in sport and physical activity, it is important that we consider some facts concerning clotting and the fibrinolytic system. Exercise has long been known to hasten the clotting process. Furthermore, the increase in epinephrine associated with the fight-or-flight response was known to stimulate exercise-related clotting changes. Hypercoagulability, or faster blood clotting, may be viewed as an asset for an active, healthy young man involved in a strenuous contact sport such as football, wrestling, or boxing. However, enhanced blood clotting could prove fatal for less active, middle-aged men, and many find it necessary to take anticoagulants to avoid the risk of stroke or a heart attack.

Thus, the age and fitness of the subject and the stress associated with the activity, must be considered as we discuss exercise and blood clotting. It is interesting to note that the factors mentioned interact, for as one continually engages in a particular activity his fitness improves, and the stress of the activity may decline. Hypercoagulability following exercise is more pronounced in unfit subjects. It is also associated with stressful activities, including situations that do not involve physical activity at all. When a golfer has a bad day we cannot blame his rage and associated hypercoagulability on the physical activity. It is obviously related to his poor performance and the way he views the game. Thus, the oft-reported decrease in clotting time associated with exercise may have been due to the physical fitness and experience of the subjects or the nature of the exercise.

Whiddon, Sharkey, and Steadman (1969) placed subjects on a treadmill training program and measured clotting time and stress. In these experiments, we found that the initial exposure to a strenuous treadmill test prompted an increase in stress hormones and a decrease in clotting time. After several weeks of training and familiarity with the treadmill and the difficulty of the test, the stress indicators returned to normal as did the clotting time. Thus, even with an increased work load the familiar test did not elicit a change in clotting time. This may help to explain why trained subjects seem to experience less of a decrease in clotting time. Their fitness and familiarity with exercise seem to condition the magnitude of the stress response and the extent of the clotting changes.

The fibrinolytic system seems to be inhibited by the adrenocorticotropic hormone (ACTH). Ganong (1971) has defined stress as any of the multitude of stimuli that result in the liberation of increased quantities of ACTH. Thus, the epinephrine activated by the effects of stress on the sympathetic nervous system hastens blood clotting while the ACTH activated by the effects of stress on the anterior pituitary gland inhibits the lysis or breakdown of clots. Both of these reactions are mediated by the hypothalmus, where our emotional response to a situation is formulated. Since we know that an intravascular clot can cause a stroke or a myocardial infarction (heart attack) we cannot ignore the effects of stressful exercise on untrained, middle-aged, coronary-prone men. We shall deal with the problem of assessing what constitutes stressful physical activity in Chapter 8.

It was once believed that physical training enhanced the fibrinolytic system. This proposition was advanced as evidence of the beneficial effect of physical activity on cardiovascular health. However, the current view is that fibrinolytic activity is enhanced immediately after vigorous activity (Iatridis and Ferguson, 1963), and that prolonged, fatiguing exertion may depress the system. Therefore, the proposition regarding physical activity and cardiovascular health must be modified to suggest that *regular, moderate* physical activity will regularly enhance fibrinolytic activity. There does not seem to be any training benefit involved, only the acute effect following exercise. (Moxley, Brakman, and Astrup, 1970.)

We cannot leave this discussion of blood clotting without a word concerning women. You are all aware of the controversy involving the undesirable side effects allegedly associated with the use of the pill. Some brands carry distinct warnings that indicate up to a twofold increase in the incidence of thromboemboletic (clotting) disorders among those using oral contraceptives. Others make no mention of such side effects. At this time it is impossible to make a definitive statement on this subject since the concentration and ratio of hormones differs in each brand or type of oral contraceptive, and the debate regarding the dangers of their use still rages in the medical literature. Our own limited findings suggest a faster clotting time for those using the pill. We have also documented the case of a young woman who had to undergo anticoagulant therapy and surgery to remove clots that appeared only 3 days after she undertook a program of exercise at a weight-control studio. Previously she had taken the pill for several years, but had not engaged in physical activity for some time. It is recommended that anyone taking the pill should consult with her physician before undertaking any strenuous and unfamiliar exercise. Moderate physical activity does not seem to significantly influence the clotting time for regularly active young women who use the pill.

THE HEART

The heart is truly an amazing muscular pump; it is the ultimate endurance muscle, beginning its rhythmic contractions long before birth and continuing its work until the moment of death. We are so intrigued by our heart that we extol its virtues in poetry and song. This gratitude is understandable since most of us have experienced fears that our pounding heart might explode, and have been filled with relief when the crisis has passed. Here we shall approach the heart with less emotion and consider its capabilities as a pump, for it is this function that is so essential to all our other activities.

Cardiac Output

Although the heart muscle has a basic rhythmicity all its own, it is typically driven at a higher rate by accelerator impulses arriving at the sino-atrial node via the sympathetic nervous system. Cholinergic stimulation via the vagus nerve of the parasympathetic nervous system slows the rate of the heart, probably by altering the permeability of the myocardial membrane. When the sino-atrial node is depolarized the depolarization wave spreads rapidly across the atria to the atrioventricular node. Depolarization then spreads down the Purkinje fibers to the ventricles. We shall see that the rate of the beat is an important factor in the capacity of the heart as a pump. This pumping capacity, or cardiac output ($\dot{Q}$), is an important factor in our adjustment to the demands of exercise. One of the major ways in which we adjust blood flow to meet the demands of exercise is through an increase in the amount of blood pumped per minute. Cardiac output is defined as the product of the heart rate (HR) times the stroke volume (SV), that is,

$$\dot{Q} = HR \times SV$$

At rest the cardiac output averages about 5 liters per minute. This blood flow is similar in both trained and untrained subjects. The untrained individual may accomplish the task with a heart rate of 72 while the trained subject may have a rate of 50 or less.

Untrained:
72×70 ml $=$ 5 liters per minute
Trained:
50×100 ml $=$ 5 liters per minute

Thus, it would appear that the trained subject would have a larger stroke volume and a lower heart rate. In fact, one of the best-documented effects of training is a reduction in the resting heart rate.

Heart rate and stroke volume do not increase in a similar fashion with an increasing work load (Fig. 6.1). As the work load and oxygen consumption increase, the heart rate increases in a linear fashion. On the other hand, the stroke volume seems to level off at low to moderate work loads. Therefore, any further increases in cardiac output must be due to the influence of the heart rate increase alone. This point is of considerable importance since it supports the use of the exercise heart rate as an indicator of the cardiac output as well as of the oxygen intake (see Fig. 6.2). This information will be utilized in Chapter 12 when we discuss the use of the heart rate in the prescription of exercise.

TRAINING

The graphic presentations in Figs. 6.1 and 6.2 clearly illustrate the effects of training on the heart rate, stroke volume, and cardiac output. The trained individual has a lower resting heart rate and a somewhat lower maximal heart rate. He has a higher

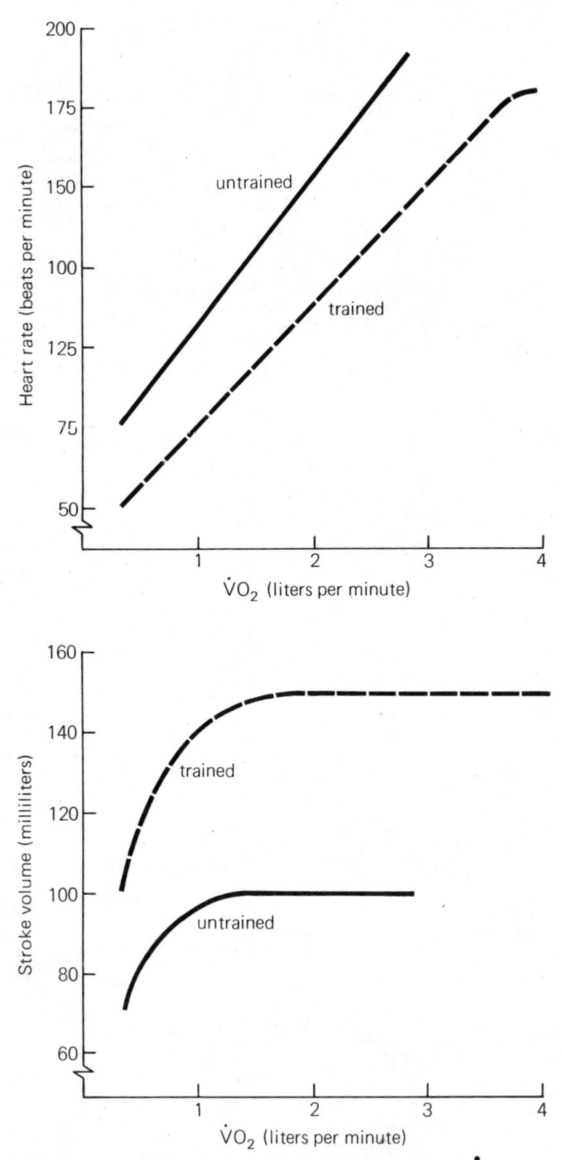

Figure 6.1. Relationship of work load ($\dot{V}O_2$) to heart rate and stroke volume. At work loads above 1.5 liters per minute, any increase in the cardiac output depends on an increase in the heart rate.

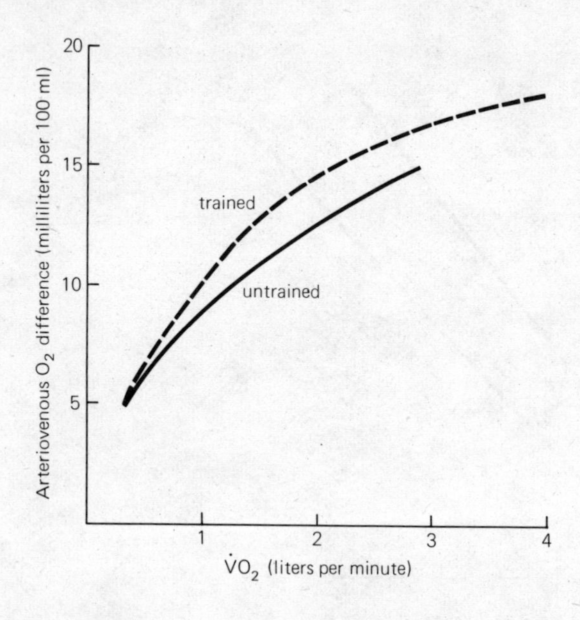

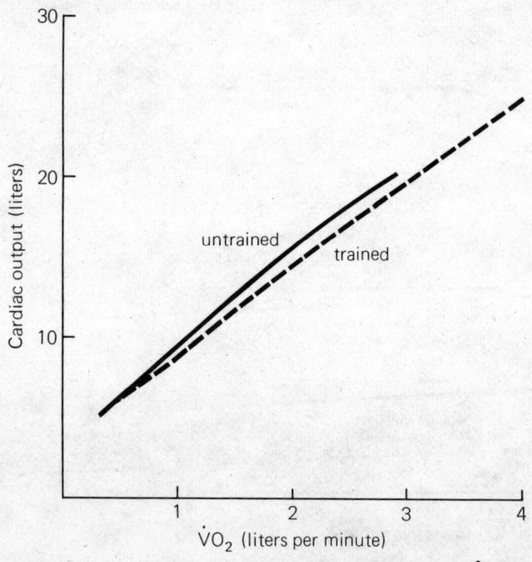

Figure 6.2. Relationship of work load ($\dot{V}O_2$) to arteriovenous oxygen difference and cardiac output. The increase in cardiac output, while almost linear, tends to flatten somewhat as the maximal stroke volume is reached (see Fig. 6.1).

resting and maximal stroke volume, with some persons having rates of 200 ml. (Ekblom and Hermansen, 1968.) It is possible for a highly trained athlete to achieve maximal cardiac output approaching 40 liters per minute, but most of us must be content with a cardiac output well below that figure. Table 6.1 illustrates some similarities and differences between a healthy but untrained individual and a superb world class endurance athlete.

We defined cardiac output as the product of heart rate and stroke volume. However, because the stroke volume is rather difficult to measure, even in laboratory animals, another equation is used to calculate $\dot{Q}$. The Fick equation utilizes easily measured oxygen-consumption data and the arteriovenous oxygen difference. The arteriovenous measurements are made using oxygenated arterial blood from a systemic artery and mixed venous blood, usually taken from the pulmonary artery. This measurement requires the insertion of a catheter into a forearm vein. The catheter is then worked into position in the pulmonary artery. Since this method involves some danger to the subject, the *indicator-dilution method* is used more frequently to determine $\dot{Q}$ in studies of physical activity. This method involves the injection of a known amount of harmless dye, after which a series of blood samples is taken to determine the concentration of dye after its initial circulation through the heart. Once the cardiac output has been determined the stroke volume can be calculated by dividing $\dot{Q}$ in liters per minute by the heart rate in beats per minute.

The effect of training on the arteriovenous oxygen difference is illustrated in Fig. 6.2. The shift of the hemoglobin saturation curve during vigorous exercise and the increased aerobic capacity of trained muscle cells combine to allow a greater extraction of oxygen from a given quantity of arterial blood. The effects of training on the factors involved in cardiac output and gas transport have yet to be fully understood. The decline in the resting heart rate as well as the lowered heart rate at a submaximal work load may be caused by the increase in stroke volume. The increase in stroke volume is the result of a more efficient ejection of blood during each beat. It appears that the untrained heart ejects only a portion of the fluid it

TABLE 6.1. A COMPARISON OF RESTING AND MAXIMAL VALUES FOR AN UNTRAINED INDIVIDUAL AND A HIGHLY TRAINED ATHLETE

Rest/Maximal

	$\dot{Q}$ (*liters/min*)	HR (*bpm*)	SV (*ml/beat*)
Untrained	5.0 / 20.0	72 / 200	70 / 100
Highly Trained	5.0 / 38.0	42 / 190	120 / 200
	Cardiac Output	*Heart Rate*	*Stroke Volume*

$$\dot{Q} = HR \times SV$$

$$\text{untrained} = 200 \times 100 = 20 \text{ liters}$$

$$\text{trained} = 190 \times 200 = 38 \text{ liters}$$

Sources: Ekblom and Hermansen, 1968; Saltin and Astrand, 1967.

contains. Training seems to increase the proportion ejected, either by an increase in the tonus of the cardiac musculature or via nervous system and hormonal mechanisms. It does seem that some of the effects of training on cardiac output may be mediated by the CNS. Recent studies have indicated that the autonomic nervous system may be subject to learning or conditioning effects. (Miller, 1969.) Human subjects have been able to control their heart rate and make it increase or decrease. It is possible that some events in a training program reinforce lower heart rates. Thus, it may be that some part of the decline in resting and submaximal heart rates may be due to learning

	Rest/Maximal	
$\dot{V}O_2$ *(liters/min)*	*Arteriovenous O_2 Difference (ml/100 ml)*	$\dot{V}$ *(liters/min)*
.250 / 3.00	4.5 / 15	6 / 120
.250 / 6.00	4.5 / 16	6 / 200
Oxygen Uptake	*Arteriovenous O_2 Difference*	*Pulmonary Ventilation*

$$\dot{Q} = \frac{\dot{V}O_2}{\text{a-v } O_2 \text{ Difference}} \quad \text{(Fick equation)}$$

$$\text{UT} = \frac{3.0}{15} = 20 \text{ liters}$$

$$\text{TR} = \frac{6.0}{16} = 38 \text{ liters}$$

or conditioning and another part due to actual changes in the stroke volume (Table 6.2).

EXERCISE AND THE CONTROL OF CARDIAC OUTPUT

When muscular contractions begin, local hypoxia (i.e., lack of oxygen) seems to bring about local vasodilation and an increase in the flow of blood to the muscles involved. Should the activity continue, the massaging action of the contractions will assist the *venous return* which is so essential for the maintenance of the stroke volume. You can see how static contractions could actually reduce venous return and stroke volume.

TABLE 6.2. THE OXYGEN PULSE (O_2-PULSE)[a]

	O_2Pulse and Workload[b]		
	Rest	*Submaximal Load*	*Maximal*
Untrained	.250 ÷ 70 = 3.5	2.1 ÷ 160 = 13.1	2.9 ÷ 190 = 15.3
Trained	.250 ÷ 50 = 5.0	2.1 ÷ 140 = 15.0	4.0 ÷ 180 = 22.2

[a] The O_2-pulse, the oxygen used per beat of the heart, is determined by dividing the oxygen used per minute by the pulse rate (e.g., 1.5 liters ÷ 150 = 10 ml per beat).

[b] The O_2-pulse is sometimes used as an index of the stroke volume since it is relatively easy to measure the $\dot{V}O_2$ and the pulse rate.

When exercise begins, the vasomotor center in the medulla receives stimuli from higher brain centers. Should the activity continue, impulses from thermo-, chemo-, and baroreceptors will also prove important. The vasomotor center sends impulses via sympathetic nervous pathways to increase the heart rate and stroke volume. Sympathetic fibers also activate (1) vasoconstriction of arterioles in order to divert blood to the working muscles, (2) venoconstriction to reduce the volume of blood pooled in the veins, and (3) vasodilation of arterioles serving the working muscles. In addition to the three- to eightfold increase possible for cardiac output, blood flow needs are also met by a *redistribution* of blood from inactive or less involved areas to the active muscles. The intensity of exercise dictates the increase in cardiac output as well as the degree of redistribution required. Some digestion may go on during light activity but the major portion of the blood will be diverted away from the digestive organs during maximal effort. (Rowell, 1971.)

Myocardial Oxygen Supply During Exercise

The coronary circulation carries oxygen and fuel to the cardiac musculature. While at rest, the heart extracts about 75 percent of the oxygen in the blood provided by the coronary circulation so that the increased oxygen needs of the myocardium during exercise must be met by an increase in coronary bloodflow. The increased blood flow depends on a dense network of capillaries, with at least one capillary per muscle fiber. Local hypoxia has a potent dilating effect on the arterioles serving the myocardial capillary network. It has been suggested that training improves capillary density (Poupa and Rakušan, 1966), and that exercise stimulates the development of so-called collateral vessels (Eckstein, 1957). Stevenson et al. (1964) reported an increase in coronary vascularization after running or swimming. It is most interesting to note that moderate exercise is more effective in this regard than is more strenuous effort. (The health implications of these studies will be discussed in Chapter 11.)

The maintenance of the myocardial oxygen supply is crucial because the heart cannot utilize anaerobic energy sources. Thus, anything that increases its oxygen needs or diminishes its blood flow may be considered potentially dangerous to this vital organ. A clot or thrombus is probably the greatest threat to the myocardium. Exercise also calls forth a considerable increase in the oxygen needs of the myocardium. This increase is related to the muscular tension that must be developed and the time it is maintained ($T \times T$). The increase in heart rate during exercise increases the time that tension is maintained per minute. The increased myocardial oxygen needs that parallel the rise in cardiac output (which is related to heart rate) can exceed a fivefold increase during maximal effort. The oxygen needs of the heart can become crucial when the coronary blood flow is compromised. However, regular moderate physical activity does not seem to compromise the oxygen supply to the healthy heart.

Because tension must increase with arterial blood pressure, and because blood pressure rises markedly during maximal lifting efforts, physiologists have long cautioned against the use of heavy resistance exercises for untrained adults. The strong static component of the lifting effort calls forth a marked rise in pressure as well as a disproportional increase in heart rate. Both of these factors increase the tension × time index and the oxygen needs of the myocardium. Moderate, rhythmic physical activity provokes but moderate increases in heart rate and blood pressure, and even seems to be tolerated well by those recovering from a heart attack (Chapter 12).

Energy for the Heart

At rest, the myocardium derives its energy primarily from free fatty acids (FFA), lactate, and glucose (about 40, 30, and 30 percent respectively). These contributions remain similar during light activity, but as exercise intensity increases, FFA and glucose metabolism decline and lactate provides as much as 60 percent of the energy required. During exercise of long duration, the lactate and glucose contributions eventually decline and the FFA utilization rises to almost 70 percent of the total energy production. Athletes seem to produce and utilize less lactate at a given work load. They seem better able to mobilize and utilize FFA as an energy source for myocardial metabolism. (Keul, 1971.)

EXERCISE AND TRAINING

It has been suggested that endurance training will lead to cardiac hypertrophy, that is, a functional increase in the size of the cardiac musculature. This proposition has been demonstrated in animals, but is still a debatable issue in regard to humans. Even if athletic subjects can be shown to have larger hearts, it is difficult to separate the role of heredity from the effects of training. Recent studies have reinvestigated the effects of exercise on cardiac weight and probed the oxidative capacities as well. While training did increase the heart weight : body weight ratio in rats, it did not alter the concentration or activity of specific oxidative enzymes. The authors concluded that the aerobic capacity of normal, untrained rat myocardium was sufficient to meet the demands of the endurance training program without the need for an adaptive increase in respiratory capacity. They also indicated that the respiratory enzyme concentration was five times higher in the heart than in skeletal muscles. (Oscai, Molé, and Holloszy, 1971.)

CIRCULATION

The focus of this monograph is the physiology of physical activity, hence, our concern in this section is with the blood flow to the muscles. We have already mentioned that blood flow to the working muscles is increased via an increase in the cardiac output, as well as by a redistribution of flow from less active regions of the body. Thus, muscles that receive but 20 percent of the resting cardiac output (0.20 × 5 liters = 1 liter) at rest may require almost 90 percent during maximal effort (0.90 × 25 liters = 22.5 liters). While cardiac output may increase some five or six times, muscle blood flow can be increased twenty times or more.

Hemodynamics

Blood flow is governed by the pressure driving the blood through the vascular system and the resistance that acts to oppose this flow. This relationship may be viewed in a simplified equation. We see that as pressure (P) increases or resistance (R) decreases the flow will increase, but as pressure (P) decreases or resistance (R) increases the flow will be diminished.

$$\text{Blood flow} = \frac{P}{R}$$

The contraction of the left ventricle forces blood into the aorta and the elastic arteries. The peak pressure is called the *systolic pressure,* and it occurs during ventricular systole. The arterial blood pressure falls during ventricular diastole to a low point, referred to as the *diastolic pressure.* The average or mean pressure throughout this cycle is the mean arterial pressure, and it serves to propel the blood through the arteries, arterioles, capillaries, venules, veins, and back to the heart. The mean pressure may average above 90 mm Hg in the aorta, drop to 10 mm Hg in the capillary network, and finally reach zero in the right atrium.

The *blood pressure* typically averages about 120 mm Hg systolic and 80 mm diastolic (120/80). During exercise involving rhythmic contractions, systolic pressure increases while diastolic pressure is relatively unchanged. During sustained static contractions, both systolic and diastolic pressures increase. (see Fig. 6.3.) The diastolic increase is due to the rise in peripheral resistance that results when the contraction restricts blood flow in the muscle. Blood pressure is affected by changes in cardiac output, peripheral resistance, arterial elasticity, blood volume, and blood viscosity. We have just noted how exercise

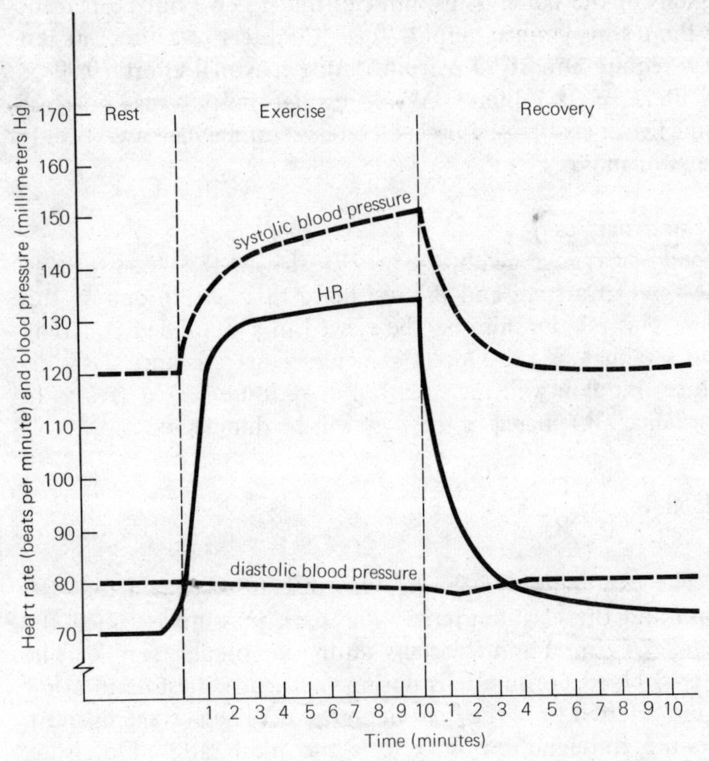

Figure 6.3. Typical heart rate and blood pressure response to a submaximal work load (such as jogging). Notice the slight rise in HR just prior to exercise, the so-called anticipatory increase. The HR and systolic blood pressure begin to level off after about two minutes of exercise. The continued rise in HR illustrates the need to increase the circulation of blood to the skin in order to dissipate the heat generated during exercise. During a rhythmic exercise (jogging, cycling), the diastolic blood pressure declines a bit or remains unchanged. During strong static contractions, the diastolic pressure increases markedly (to 100 mm Hg and beyond). Immediately after exercise, the systolic blood pressure declines quickly toward resting levels while the HR remains elevated for some time to help remove waste products and to supply the post-exercise demands for oxygen and fuel. The diastolic pressure sometimes exhibits a post-exercise depression lasting up to a few minutes. Fit individuals demonstrate less dramatic changes in HR and blood pressure (for the same work load) and a more rapid return to resting values.

alters the cardiac output and peripheral resistance. Arterial elasticity is certainly affected by age. Blood volume may be increased after training or transfusion, and be decreased via hemorrhage. The viscosity of the blood is altered during exercise because the increase in blood pressure seems to force fluid from the capillaries. However, this tendency is eventually balanced by the osmotic influence of the blood proteins. We will touch on the problem of high blood pressure (hypertension) when we deal with the implications of physical activity and cardiovascular health (see Chapter 11).

It seems obvious that blood flow will diminish as *resistance* to flow is increased. Flowing blood loses energy as it travels through a tube because the friction of fluid flowing over vessel walls must be overcome. This friction depends on the viscosity of the fluid, the length and diameter of the tube. While the length of the blood vessel is not likely to change, the viscosity or thickness of the blood can increase when the number of red blood cells rises from 5 million (per cubic millimeter) to 6 million or more during acclimatization to altitude. The importance of arteriole diameter has also received attention in relation to the redistribution of blood flow. Vasoconstriction increases resistance while vasodilation lowers the resistance to flow. During exercise the vasoconstriction in inactive organs is more than balanced by vasodilation in the active muscles.

Microcirculation

The peripheral blood vessels (arterioles, metarterioles, capillaries, and venules) can be viewed as a well-organized microcirculation designed to serve several functions. In addition to the obvious function—the return of blood to the heart via venules and veins—the peripheral vessels also serve to maintain blood pressure with peripheral resistance provided by arterioles and, to some extent, the metarterioles (the smaller, less muscular arterioles). Blood flow is distributed to the active tissue or returned via the venules due to the local control of the metarterioles and the precapillary sphincters. A final portion of the microcirculation, the arteriovenous anastomoses, allow a complete bypass of the true capillaries as they provide a direct route from arterioles to venules. It is likely that the arteriovenous anastomoses are responsible for temperature control, directing heat to surface veins and away from the capillaries located deep in the tissue. Their muscular walls are under sympathetic nervous control and, when the body temperature rises, the anastomoses can divert needed blood away from the true capillaries. (See Chapter 7 for a further discussion of thermoregulatory mechanisms.)

Capillaries We have already mentioned that capillaries are more profuse in the region of red muscle fibers. This observation suggests the possibility that training may either increase the capilarization of the muscle, or that training may somehow enhance utilization of the available capillaries. Hermansen and Wachtlová (1971) studied the capillary density of skeletal muscle in trained and untrained men. Muscle biopsies from the highly trained subjects (Max $\dot{V}O_2 = 71.4$ ml per kilogram of body weight per minute) did not reveal differences in the number of capillaries per square millimeter when compared with the less well trained (Max $\dot{V}O_2 = 50.2$ ml), with values of 640 and 600 per square millimeter respectively. However, because the average size of the muscle fiber was 30 percent larger in the highly trained subjects, the ratio of capillaries to muscle fibers was significantly higher for the trained subjects (1.49 versus 1.08 capillaries per muscle fiber). Hermansen and Wachtlová concluded that the primary effect of endurance training seemed to be an increase in the size of muscle cells with a secondary increase occurring in the number of capillaries per fiber.

Using similar techniques, Pařízková et al. (1971) studied the capillary : fiber ratio in young and old men. They found that the young men had significantly higher capillary : fiber ratios, and that among the old men (over 73 years of age), the ratios did not differ for active or inactive subjects. Thus, it appears that training and physical activity may enhance capilarization, but that the benefits may eventually be obscured by the process of aging. These studies support the benefit of training on the capillary : fiber ratio, but do not support earlier claims that training increased the number of capillaries per unit surface area of muscle (per square millimeter). It seems likely that the diffusion distance from capillary to muscle tissue is not a limiting factor in oxygen transport and physical performance, except perhaps during senescence.

CARDIORESPIRATORY ENDURANCE

We have been discussing the involvement of the cardiovascular and respiratory apparatus during exercise, as well as the effects of chronic exercise (i.e., training) on these systems. These systems work together to provide for our oxygen needs, to transport foodstuffs and wastes, and to assist in the maintenance of body temperature and acid-base balance. The capacity of the so-called cardiorespiratory system—the aerobic capacity or maximal oxygen intake—serves as the best single measurement of endurance fitness. Because of its relationship with various respiratory and cardiovascular functions it also serves as a measure of the fitness of these systems. The aerobic capacity indicates the ability to *take in*, *transport*, and *utilize* oxygen in the working muscles.

Aerobic capacity is related to performance in endurance activities as well as in vigorous physical work. However, oxygen intake alone does not predict the winner in a sporting event or the best performance in industrial tasks. As Cumming (1971) stated, "The exercise physiologist is virtually unable to predict who is going to do well in a given sporting event from basic laboratory tests." Skill, motivation, and other factors operate to confound such predictions. On the other hand, aerobic capacity can be improved with regular physical activity, the sort of moderate physical activity that leads to the health benefits discussed in Chapter 11. We view cardiorespiratory endurance as the *best single measure of physical fitness* because it incorporates both the health and the athletic connotations implied by the phrase. We do not ignore the implications of athletic fitness in the lives of young men and women, nor should we ignore the contribution of physical fitness to the health and quality of adult lives.

CHAPTER 7 ENVIRONMENTAL FACTORS

We have included pertinent environmental factors in Part II because of their interaction with the supply and support systems. Although a thorough discussion of each is beyond the scope of this monograph, we shall attempt to highlight the influences of temperature, altitude, pressure, and air pollution on physical activity and maximal performance capacity.

TEMPERATURE REGULATION

At rest, our metabolic heat production amounts to about 1.2 kcal per minute or 72 kcal per hour. Moderate exercise can elevate that heat production to 600 kcal per hour or more. Thus, you can see that exercise by itself can create considerable heat. When exercise is performed in a hot humid environment the metabolic heat cannot be dissipated and the body temperature rises. The problems associated with exercise in hot environments are several. *Heat cramps* occur when considerable salt is lost in the perspiration, but they can be avoided by the judicious use of salt in the diet. *Heat exhaustion* occurs when the heat stress exceeds the capacity of the homeostatic mechanisms. The person with cold skin, a weak pulse, and a dizzy feeling should be given fluids and allowed to rest. *Heat stroke* is the result of the failure of the thermoregulatory system. This severe heat disorder can lead to excessive water loss, extreme body temperature, electrolyte imbalance, and, possibly, death. Such disasters can be avoided with adequate hydration and by the cessation of physical activity when high temperature or humidity combine to overload the thermoregulatory mechanisms.

The temperature-regulating mechanism, located in the hypothalmus, acts like a thermostat to maintain the body temperature at or near the so-called normal (37 C or 98.6 F) level. The regulatory center responds to the temperature of the blood perfusing the hypothalamus. If the blood becomes cooler the thermostat sends information to conserve heat loss via vasoconstriction of cutaneous blood vessels. Heat can also be generated by shivering. If the blood temperature rises above the desired level the regulatory center can affect vasodilation of blood vessels in the skin and stimulate the production of sweat. Consequently, blood is brought from the warmer core to the surface, allowing heat loss via conduction, convection, and radiation, as well as by the evaporation of sweat from the surface of the skin.

Cutaneous heat and cold receptors also aid in the maintenance of body temperature. Stimulation of the cold receptors causes the information to be relayed to the sensory cortex as well as to the thermoregulatory center in the hypothalmus. These impulses may reset the thermostat at a higher level, activate heat-conserving measures, and also cause us to consciously seek relief from the cold. Similarly, a hot environment will cause a lowering of the thermostat's set point, the activation of temperature-lowering reflexes, and the desire to seek relief from the heat. The cold of the ski slopes will result in the vasoconstriction of cutaneous blood vessels, especially in the extremities. The resulting discomfort can best be overcome by the elevation of the body (core) temperature via vigorous physical activity. We can also put on warmer clothing or seek relief in the lodge.

Exercise in the Heat

Exercise in the heat poses more difficult problems for our thermoregulatory center. When we begin to exercise, our regulatory center increases the thermostatic set point and our body temperature is allowed to increase. In a moderate environment, the rise in body temperature will depend on the relative work load, that is, the percentage the exercise oxygen consumption represents of the maximal oxygen intake. The temperature will increase about one degree at 50 percent of work capacity (38 C), and will rise above 39 C at the maximal oxygen intake level. Thus, in the fever of maximal effort the temperature may rise above 102 F, even in a cool environment. This resetting of the core temperature during exercise could be viewed as an adjustment favorable to the enzyme activity within the muscles. Under moderate environmental conditions the various methods of heat dissipation are not employed until the elevated set point has been reached.

In hot or humid environments we are able to maintain temporary thermal balance during exercise by virtue of circulatory adjustments and the evaporation of sweat. In a hot dry environment the body actually gains heat when the air temperature exceeds the temperature of the skin. Under these conditions the evaporation of sweat allows the maintenance of thermal equilibrium. However, when the humidity is also high and *evaporation cannot take place,* the body temperature continues to rise and physical performance is severely impared. The circulatory response to heat diverts a significant portion of the cardiac output from the working muscles to the cutaneous blood vessels. The result is a higher exercise heart rate for a given work load and a reduction in work capacity. Similarly, the high sweat rates elicited by exercise in the heat (above 3 liters per hour) can lead to dehydration, reduction of blood volume, and to an even greater circulatory distress, to say nothing about the loss of needed electrolytes (salts) and the alarming rise in body temperature (hyperthermia).

Sweating The total body fluids amount to about 40 liters (for a 70-kg man), of which 15 liters are located outside of the cells (extracellular). In a normal day we lose, and must replace, about 2.5 liters of water. Of this amount, about 0.7 liter is lost from the skin and lungs (insensible water loss), 1.5 liters from the urine, 0.2 liter from the feces, and about 0.1 liter via perspiration. During heavy exercise in the heat, the water loss via sweating can be increased beyond 3 liters per hour. The sweat produced in the eccrine glands of the skin may amount to 12 liters per day of a dilute sodium chloride (NaCl) solution (often 0.4 percent, Folk, 1974). Work capacity becomes compromised as the water loss progresses. Therefore, it is essential that the fluid be replaced if performance is to continue. Greenleaf and Castle (1971) studied the effect of hypo and hyperhydration on the rectal (core) and skin temperatures of men during bicycle ergometer exercise (50 percent of Max $\dot{V}O_2$). Rectal and skin temperatures were significantly higher with 5 percent dehydration. During hyperhydration the rectal temperature remained below the value measured during ad lib water replacement. Thus, it seems that sweating rates and core temperatures depend upon adequate water replacement, and that our thirst tends to underestimate water needs during exercise. Hyperhydration increases the capacity for evaporative heat loss and, therefore, can reduce the load on the circulatory system. Consequently, it is not surprising that hyperhydration has led to increased work performance in hot industrial environments.

Salt loss Water replacement alone will not compensate for the loss of sodium, chloride, and potassium in the sweat. Most authorities agree that these needs can be met by the liberal use of salt during meals. For prolonged periods of work in the heat, an electrolyte solution should be provided. In a 1971 news release the National Collegiate Athletic Association (N.C.A.A.) Committee on Competitive Safeguards and Medical Aspects of Sports recommended solutions containing potassium as well as sodium chloride. Solutions containing the necessary electrolytes, as well as glucose, can be obtained commercially. An inexpensive solution can be prepared by adding salt to frozen citrus fruit-juice preparations (about 1 teaspoon per quart) or by replacing half of your fluid needs with tomato juice and the balance with water.

It is interesting to note that the traditions and rules of sport often operate in conflict with sound physiological principles. The restriction of water during football practice continues to result in tragic and avoidable deaths from heat stroke and heat exhaustion (eight deaths in 1970). The rules governing international competition in the marathon prohibit fluid administration until the marathoner has completed 10 km of the 42-km event (26 miles, 385 yards). Costill, Kammer, and Fisher (1970) reported that the combination of limited gastric emptying, rapid fluid loss, and the international rules severely limit the athlete's ability to replace the fluids lost *during* the event. However, despite the acute dehydration, these superbly trained athletes seem able to maintain their blood volume at the expense of the intracellular fluids. Thus, marathoners are able to maintain circulation to the working muscles while undergoing an acute fluid loss that would lead to circulatory embarrassment in untrained subjects. Costill, Saltin, Söderberg, and Jansson (1973) recommend a cool electrolyte solution with a low concentration of glucose to insure gastrointestinal absorption *during* prolonged exercise.

Heat Stress

As you may have guessed, heat stress, or the effect of heat on the body, cannot be predicted on the basis of the air temperature alone. The relative humidity is an important factor determining the effect of sweating on evaporative heat loss. If the sweat cannot evaporate, the sweating mechanism only adds to the thermal stress by virtue of fluid loss. Air movement and radiant heat are also important factors to consider when evaluating the effects of a given environment on human comfort and performance. Finally, the metabolic heat production due to physical activity must be considered along with the type and color of clothing worn during the exposure. The WBGT (wet bulb globe temperature) heat-stress index provides a simple and accurate indication of the effect of the environmental factors on active humans. The judicious use of such an index, along with the enlightened provision of fluids and electrolytes during physical activity, could virtually eliminate the tragic loss of life due to heat exhaustion and heat stroke.

The WBGT index utilizes the dry and wet bulb temperatures to assess air temperature and relative humidity. The black copper globe indicates radiant heat as well as air movement. The several temperatures are weighted to indicate their contribution to the total heat stress (wet bulb $\times$ 0.7 + dry bulb $\times$ 0.1 + black globe $\times$ 0.2 = WBGT in degrees F). The Marine Corps suggests the utilization of discretion when the index exceeds 80 F, the avoidance of strenuous activity in the sun at values above 85 F, and the cessation of physical training when the index exceeds 88 F. Trained individuals who have been acclimatized to the heat are allowed to continue limited activity at index temperatures above 88 F. Any school able to afford a football helmet can afford to construct and utilize this simple index. It should be noted that the index was originally developed for military use and is based on data gathered using ordinary military fatigue uniforms. Football uniforms, especially those with dark jerseys and helmets, will add to the radiant heat problem and diminish evaporative cooling as well. A modern version of the WBGT, the WGT wet globe thermometer index (Fig. 7.1), provides similar environmental information in a compact unit without the need for computing the final index value.

Acclimatization

On the first day of exercise in a hot environment, a worker or athlete will experience a near maximal heart rate, elevated skin and core temperatures, and severe sensations of fatigue. Within several days of similar exposure to work in the heat, the task is accomplished with a reduced heart rate made possible by an improved blood distribution and an increased blood volume. Cutaneous circulation and heat conductance improve, and a gradual improvement in evaporative cooling results. The loss of water in the urine diminishes, and the salt concentration

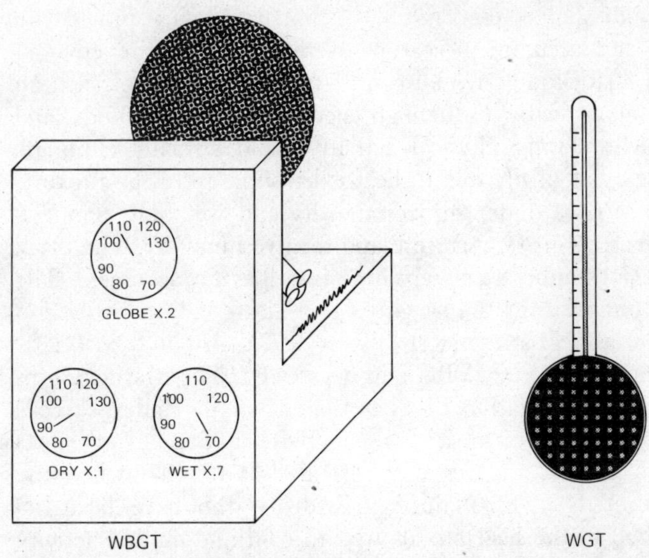

Figure 7.1. Heat-stress indicators. The WBGT index utilizes three thermometers while the WGT index is derived from one reading. The WGT thermometer is encased in a hollow black globe that is covered by a moistened webbing.

of the sweat is gradually reduced. This increase in circulatory and evaporative cooling efficiency is termed *heat acclimatization,* and it usually occurs after 4–8 days of work in the heat. (Robinson, 1967.)

Since we have already mentioned the effect of physical training on circulatory efficiency, it should not come as a surprise to learn that fit individuals are better able to work in the heat. However, since acclimatization is essentially a peripheral phenomenon, even the fit must submit to the process. It appears that the popular interval-training techniques utilized by track and swimming coaches are effective as preconditioners for work in the heat. The interval training is partially effective, even when conducted in a cool environment. Thus, it would help to prepare already fit young men from a cool climate for the demands of athletic competition in a warmer climate. (Robinson, 1967.) To be effective as a stimulus to acclimatization, exercise in a cool environment must be sufficiently intense to prompt a high core (rectal) temperature. Because exercise of this sort is too demanding for less fit individuals, they would be wise to acclimate via periods of light to moderate activity in a hot environment, alternated with rest periods and the ingestion of water and electrolytes.

Exercise in the Cold

Because of the metabolic heat generated during exercise, cold temperatures do not pose a threat similar to that imposed by hot humid conditions. But severe exposure to low temperatures and high winds can lead to frostbite, freezing, hypothermia, and even death (see Table 7.1). Peripheral vasoconstriction increases the insulating capacity of the skin, but it also results in a marked reduction in the temperature of the extremitics. This protective vasoconstriction often leads to severe discomfort in the fingers and toes. To relieve the discomfort it is necessary to raise the core temperature to allow the reflexive return of blood to the peripheral tissues. While shivering may cause some increase in metabolic heat, gross muscular activity will be far more effective in restoring heat to the troubled area.

Recent advances in the quality and availability of effective winter clothing have made it possible for most of us to enjoy exhilarating winter sports. Heavy and bulky winter garments have been replaced by light but durable outfits with high insulating values (clo values). Light, synthetic waffle-weave fabrics provide an insulating barrier of air between layers of windproof nylon. Though somewhat less effective than the ultimate winter garment made of goose down, these modern fabrics are often less expensive and are easier to care for. Care should be taken to avoid fabrics that restrict the passage of water vapor because these garments invite the condensation of evaporated perspiration. This condensation can ruin a long ski tour or make your night in that waterproof sleeping bag one you'll want to forget.

While it provides protection, clothing often provides the major restriction to effective performance in the cold. The hands often cool sufficiently to limit finger dexterity although the rest of the body is comfortably warm. Gloves with sufficient insulation to maintain circulation in the hands are too bulky to allow dexterous use of the fingers. There is some evidence of racial

TABLE 7.1. WIND CHILL INDEX[a]

Wind Speed (in mph)	*Actual Thermometer Reading (in F)*					
	50	*40*	*30*	*20*	*10*	*0*
	Equivalent temperature (in F)					
Calm	50	40	30	20	10	0
5	48	37	27	16	6	−5
10	40	28	16	4	−9	−21
15	36	22	9	−5	−18	−36
20	32	18	4	−10	−25	−39
25	30	16	0	−15	−29	−44
30	28	13	−2	−18	−33	−48
35	27	11	−4	−20	−35	−49
40[b]	26	10	−6	−21	−37	−53
			Little danger (for properly clothed person)		*Increasing danger*	

[a] The chill factor illustrates the effect of wind speed on heat loss. A ten-degree reading is equivalent to −25 when the wind speed is 20 mph.

[b] Wind speeds greater than 40 mph have little additional effect.

differences in cold tolerance. In finger-cooling studies, finger skin temperatures were lower in black subjects, and cold injuries of the hands and feet were more frequent and more severe among blacks than whites during the Korean conflict. (Henschel, 1971.) Although women do not seem to tolerate heat as well as men, it is not clear that they are able to tolerate cold to a greater extent. Differences in total body fat and fat deposition could account for some of the differences in cold tolerance among individuals.

Acclimatization Are we able to adjust to the cold as we are able to acclimate to the heat? If so, what are the physiological mechanisms involved? The answers to these questions are complicated because it is difficult to separate functional changes, occurring as a result of exposure to a complex of environmental stimuli (acclimatization), from changes due to ex-

Actual Thermometer Reading (in F)					
−10	*−20*	*−30*	*−40*	*−50*	*−60*
Equivalent temperature (in F)					
−10	−20	−30	−40	−50	−60
−15	−26	−36	−47	−57	−68
−33	−46	−58	−70	−83	−95
−45	−58	−72	−85	−99	−112
−53	−67	−82	−96	−110	−124
−59	−74	−88	−104	−118	−133
−63	−79	−94	−109	−125	−140
−67	−82	−98	−113	−129	−145
−69	−85	−100	−116	−132	−148
Increasing danger				*Great danger*	

Danger from Freezing of Exposed Flesh

posure to a single environmental factor (acclimation), or a diminution in sensation associated with repeated exposure to specific environmental stimuli (habituation). However, specific examples of cold acclimatization do appear in the research literature. One example is a metabolic adjustment wherein the metabolic rate is increased as much as 35 percent. The subjects for this experiment, women divers (or *Ama*) also developed an improved tissue insulation during the winter months, when the water temperature in the Korean peninsula falls to 50 F. The Australian aborigines adapt to cold conditions with a hypothermic response, that is, they respond with a lowering of the core temperature to a more economic level (as low as 95 F). Evidence also suggests the adaptive value of a large body mass, short extremities, high levels of body fat, and a deep routing of venous circulation. (Folk, 1974.) It is clear that natural selection and heredity can play important roles in the adaptation to

TABLE 7.2. ALTITUDE AND OXYGEN

Altitude (feet)	*Barometric Pressure (mm Hg)*	*pO_2 in Air (mm Hg)*
0	760	159
3,200	680	142
6,500	600	125
10,000	523	111
14,100	450	94
18,400	380	75
23,000	305	64
29,500	230	48

Sources: Folk, 1974; Roth, 1968.

cold environments. However, it also seems likely that repeated cold exposures can lead to physiological and psychological adjustments that allow one to tolerate and enjoy physical activity in cold environments.

EXERCISE AT ALTITUDE

Sometimes it seems unnecessary to worry about the effects of altitude on human performance capacity. Why make such a fuss and carry out so much research if a few athletes decide to compete at higher altitudes, such as Mexico City (7347 feet above sea level)? Over 40 million people live at altitudes above 10,000 feet and some live at or above 17,000 feet in the Andes. However, no permanent habitations are found above 18,000 feet, indicating that such an elevation may be incompatible with adaptation and survival. (Buskirk, 1971.) As altitude increases, the barometric pressure declines and the atmospheric and alveolar pO_2 drop. When this occurs, the arterial blood is unable to become highly saturated and the tissues must operate with a reduced oxygen supply. Hence, in spite of the various adjustments made by the cardiorespiratory systems, the effect of the reduced pO_2 is always a reduction in aerobic capacity (Table 7.2).

During acute altitude exposure we attempt to adjust with a

pO_2 in Alveoli (mm Hg)	*Arterial O_2 Saturation (percent)*	*Maximum $\dot{V}O_2$ (percentage of sea level)*
105	97	100
94	96	
78	94	90
62	90	
51	86	75
42	80	
31	63	50
19	30	

greater cardiac output for a given (submaximal) work load. The heart rate is higher but the stroke volume may be lower because of myocardial hypoxia. Pulmonary ventilation increases and this hyperventilation leads to increased carbon dioxide exhalation and the acid-base disturbances associated with mountain sickness. The result of these adjustments is an impairment in work capacity, as well as a decline in our motivation to perform arduous physical activity. Does a high level of physical fitness provide some advantage to the newcomer at a higher altitude? Upon arrival at a higher altitude the conditioned individual maintains the sea-level advantage over the unconditioned subject, but no more. The trained individual will be able to do less than he could at sea level. He is just as likely to suffer from mountain sickness and he does not seem to acclimatize more readily.

Acclimatization to Higher Altitudes

Profound changes occur soon after one moves to a higher elevation, and such changes seem related to the altitude. Pulmonary ventilation is increased, but the energy cost of that effort is not thought to be a burden due to the reduced density of the air. Oxygen transport seems enhanced via increases in red blood cells, hemoglobin, and blood volume. Above 15,000 feet, red cells increase from 5 million (per cubic millimeter) at sea level to 6.6 million, while hemoglobin rises from 15 gm (per 100 ml) to over 20 gm. The increased viscosity associated with the rise in red cells does not seem to pose a problem since the hypoxia of altitude serves to relax (vasodilate) the arterioles. Higher altitude exposure may also prompt a permanent dilation or an increase in the number of lung and muscle capillaries. Finally, myoglobin, the hemoglobinlike molecule that serves to store oxygen in the muscles seems to increase with exercise at higher altitudes. (Balke, 1968.) These changes are reversible, however, and return to sea-level values within 1 or 2 weeks. How, then, is it possible for one to prepare for the demands of exercise at altitude, and how long does it take to develop the adjustments associated with altitude acclimatization?

Altitude Training

Altitude should not affect performances in short-duration events, be they running, alpine skiing, or throwing. In fact, the reduced air density may allow projectiles or bodies to move with less resistance. Thus, the bulk of the research has dealt with endurance performances that rely heavily on oxygen intake, transport, and utilization. Training for these events has been studied at various altitudes, intensities, and durations of effort. Balke, Daniels, and Faulkner (1967) suggested that the intensity of training may be more important than the altitude at which the training is accomplished, since athletes tend to reduce their pace as an adjustment to the demands of exercise at higher altitudes. They recommend that speed work be emphasized during training at higher altitudes in order that muscle power be maintained. To accomplish such high-intensity effort it may be necessary to reduce the duration of the workout or the length of high-speed anaerobic intervals. Recovery periods between such intervals of necessity will be longer. Beyond the need for high-speed work, the usual principles of training seem to apply, and no one has yet discovered any crucial time period or particularly effective combination of altitude and sea-level training. Some changes occur quickly. Red blood cell counts have risen to as high as 8.7 million (per cubic millimeter) in mountain climbers after several days above 23,000 feet. Balke,

Daniels, and Faulkner also suggested alternating periods of training at altitude and sea level to combine the physiological advantages of altitude on the aerobic capacity with those of sea level on speed and anaerobic power. Whatever the case, it is clear that the reduction in the oxygen-intake capacity at altitude will lower the point at which anaerobic metabolism occurs. In one study, maximal oxygen-intake capacities at sea level were reduced 10–14 percent at 7500 feet, but altitude training only reduced the deficit to 7–9 percent. (Balke, Daniels, and Faulkner, 1967.) It may be that the greatest effects of altitude training include the increased tolerance to the discomfort of anaerobic effort and the increased ability to work at a high percentage of the maximal oxygen intake.

In a paper published prior to the Mexico City Olympics, Balke (1968) predicted that no new records would be set in the middle- and long-distance events. Furthermore, he stated the need for altitude training for contestants entered in events lasting more than 2 minutes. He pointed out that the benefits of such training are usually seen in 2–3 weeks. A report of the track and field and swimming results from the 1968 games (Stiles, 1969) confirmed Balke's prediction. No world records were set in events that lasted much more than 2 minutes. The women's 800-meter run was won in the world-record time 2:00.8 while the men's 800-meter was also a record at 1:44.3. Keino of Kenya ran the 1500-meter event in a remarkable 3:34.9, just 0.8 percent above the record at that time. Since Keino lives and trains above 7000 feet, his performance led many to speculate that ultimate performances in endurance events might be achieved through the combination of arduous physical training in a hypoxic atmosphere (altitude).

Ultimate Training Stimulus

If tissue hypoxia is the stimulus that results in the alterations in aerobic capabilities due to exercise training, would not the added burden of atmospheric hypoxia work synergistically to further enhance the training effect? Perhaps not, since arduous effort at sea level already leads to low levels of oxygen tension in the muscles. However, several authors have suggested possible synergistic effects. But as Buskirk (1971) points out, the experiments do not adequately separate altitude effects from training effects. Dill and Adams (1971) tested the maximal oxygen intake of six high-quality, middle-distance runners (having an average age of 17.1 years) in a prealtitude test, during 17 days of training at 10,000 feet and again upon their return to the prealtitude test site. The postaltitude maximal oxygen-intake values averaged 4.2 percent above the prealtitude levels. However, no control subjects were used to test the effects of training alone on the aerobic capacity. Convincing experimental evidence of a synergistic training effect has yet to be reported, but Balke (1968) remains optimistic regarding the achievement of ultimate performance capacity via a sophisticated altitude training program.

HIGH PRESSURE

Physical activity is frequently conducted under conditions of high pressure, either in air or under water. Men work in deep mines, in pressurized caissons, or in operating rooms. They use scuba gear or pressurized suits for depths beyond 1000 feet. One could even ski at 10,000 feet in Colorado, grab a jet and cruise above 35,000 feet in a pressurized cabin, and deplane for some underwater fun in the Bahamas. Passage from low to high pressure may be accomplished rather quickly and without danger, as in the case of the free-falling parachutist. However, passage from high to low pressures must be accomplished with extreme caution.

An increase in pressure will decrease the volume of a gas without appreciably affecting the volume of water. At sea level, air exerts 14.7 pounds per square inch (psi) of pressure. As one descends below sea level the weight of the atmosphere above increases the pressure. Also, as one descends below the surface of the sea additional pressure is exerted by the weight of the water above. At 33 feet, the water alone will exert a pressure of 1 atm (14.7 psi). Since the density of water remains the same regardless of the depth (64 pounds per cubic foot), you can see that the pressure will continue to increase with depth. The pressure environment is most dangerous at the end of the period of exposure. However, during the exposure carbon dioxide buildup, oxygen needs, and the pressure itself pose problems for the underwater diver or pressurized worker. Should a diver breathe deeply from his tank at a modest depth and then hold that breath while surfacing, the once-compressed gas may expand and cause alveolar rupture, pulmonary hemorrhage, and even air emboli that may reach the heart or brain and cause death.

The length of exposure to pressure is important both for its relationship to oxygen poisoning and to the time required for eventual decompression. Oxygen poisoning results when pure oxygen is used beyond a depth of 30 feet for any length of time. The tissues become saturated with oxygen and convulsions may result. Using compressed air, a diver may descend to 100 feet or beyond and return without the need for decompression. However, should he remain submerged for any length of time the pressurized gas will force nitrogen into solution in his blood. Failure to decompress slowly and let the nitrogen pass out of the blood results in the painful condition called the *bends*.

A complete discussion of the physiology of pressure is far beyond the scope of this monograph. It could be that cold acclimatization would be helpful to the diver, but little has been written regarding possible acclimatization to the pressures of diving. The interested sportsman should consult an authoritative reference for further information about this fascinating, but potentially dangerous, activity.

AIR POLLUTION

It was a warm humid day in September and thousands of cars hurried along the highways that circled the suburban communities outside of New York City. The haze created by the action of sunlight on the hydrocarbon emissions hung heavily in the quiet air. When the Quibbletown, New Jersey, football squad took the field at 2:45 in the afternoon the temperature was 85 F and the humidity was above 70 percent. Such were the conditions in that pleasant, middle-class community (as described by Jackson, 1971). Early in the practice, players began to complain of troubled breathing, of chest pains, tightness, nausea, and some began to vomit. The scene was repeated at other area schools where young, healthy athletes engaged in vigorous physical activity were learning firsthand about the effects of air pollution. Adults were also affected as they attempted to mow their lawns or work in their gardens. The urban East had experienced the choking pall that forces Los Angeles school children to remain indoors during recess on days when the photochemical smog is particularly bad.

The ozone created by this photochemical smog, due largely to the emissions from automobiles, has been shown to create symptoms which are accentuated during exercise (Bates, 1972). It is not the same sort of pollutant that brought death to Donora, Pennsylvania, or to London, England. Nor is it the same as the airborne lead from the combustion of leaded gasoline or from smelter operations. There are many sources of air pollution and we are slowly beginning to recognize them as threats to the quality of life and to life itself.

Listed below are some pollutants and their biological effects:

Reducing agents: irritate conducting airways (bronchial tubes)
Oxidants: affect diffusing surfaces (e.g., alveolar breakdown in emphysema)
Carbon monoxide: reduces oxygen transport capacity (competes for space on hemoglobin molecule)
Carcinogens: cause cancer

Some pollutants are harmless by themselves, but in combination with other factors they are capable of exerting potent biological effects (synergistic effect). We have long recognized the effect of pollution via cigarette smoking on pulmonary function. These effects far outweigh those of industrial and urban air pollution. Heavy smoking can lead to chronic obstructive lung disease and lung cancer. The chronic effects on the airways of the young resting smoker are small. However, during vigorous exercise, the smoking of at least two cigarettes before the test increases the oxygen cost of breathing. Even temporary abstinence from cigarettes improves the respiratory picture during near-maximal treadmill exercise (Rode and Shephard, 1971). Cigarettes, automobile and industrial emissions, even airborne dust, are capable of affecting pulmonary function. Chronic exposure to some pollutants can lead to long-term disorders like asthma or even to permanent disability (e.g., black lung, emphysema). All of these pollutants threaten the conduct of physical activity and sport. While air pollution does not typically strike down healthy young athletes before it affects others, the increased pulmonary ventilation of vigorous physical activity increases the flow of pollutants into the respiratory passages. Let us hope that our outdoor activities need never be regulated in accordance with the air pollution index, and that our enjoyment of physical activity need never be compromised by man's mistreatment of the environment.

CHAPTER 8 HORMONES AND EXERCISE

In this chapter we will consider the effects of exercise on the endocrine glands, as well as the relationship of glandular secretions to physical activity. The hormones of the endocrine system function to provide metabolic controls that complement the homeostatic regulations of the nervous system.

THE ANTERIOR PITUITARY

The anterior pituitary, or master gland, is actually under the direct control of the hypothalamus. The major tropic or stimulating hormones of the anterior pituitary gland are secreted upon command of various releasing factors that originate in the hypothalamus and travel to the anterior pituitary via the portal hypophyseal vessels. One of these, the growth hormone-releasing factor, stimulates the release of *growth hormone*, the hormone associated with bone growth, increased protein synthesis, and an interesting antiinsulin effect in muscles.

The elevations of growth hormone levels, associated with prolonged exercise, may be due to the exercise itself or to the hypoglycemia or low blood sugar related to the exercise. The growth hormone seems to be partially responsible for the process of fat mobilization during exercise. The administration of carbohydrate suppresses both the secretion of growth hormone and its fat mobilization property. (Hunter, Fonseka, and Passmore, 1965.) Growth hormone may also aid the oxidation of fat by blocking the entry of glucose into the muscles. Thus, fat would be used in preference to glucose, allowing the glucose to serve the needs of the nervous system. The higher levels of blood glucose might also stimulate an increase in insulin secretion and an increase in protein formation. (Ganong, 1971.)

Thyrotropin-releasing factor from the hypothalamus stimulates the release of the thyroid-stimulating hormone from the pituitary gland. The thyrotropic hormone travels to the thyroid gland to stimulate the production of *thyroxine*. Thyroxine then travels to its site of action (i.e., target organ) by hitching a ride on the plasma proteins. Thyroxine increases the oxygen consumption of most metabolically active tissues. In fact, thyroxine secretion is increased upon exposure to a cold environment. Secondary effects include an increased catabolism (i.e., breakdown) of fat and protein and an increased carbohydrate absorption and catabolism. It also enhances the actions of epinephrine, norepinephrine, and growth hormone. Recent evidence suggests that thyroxine utilization is increased because of exercise, and there does not seem to be any evidence of exercise-induced thyroxine depletion. (Simonson, 1971.) Stress appears to inhibit thyroid secretion, either by a hypothalamic inhibition of thyrotropic hormone, or by the effect of vasoconstriction on the thyroid gland (Ganong, 1971).

Several other hormones of the anterior pituitary gland do not seem to be closely related to physical activity. These include *follicle-stimulating hormone, luteinizing hormone,* and *prolactin*. The principal hormones of the posterior pituitary gland also seem to be of little interest. *Vasopressin,* or antidiuretic hormone, is sensitive to changes in the osmotic pressure of the plasma. Like many of the hormones, it exerts its effect on the target organ, in this case, the distal tubules and collecting ducts of the kidneys, via the activation of cyclic *adenosine monophosphate* (AMP) in the cell membrane. The cyclic AMP then increases the permeability of the membrane, and water is retained. There is some evidence that daily exhaustive exercise may increase the antidiuretic activity. *Oxytocin,* the other important hormone of the posterior pituitary, does not seem to be germane to this discussion.

Adrenocorticotropic hormone (ACTH) is secreted by the anterior pituitary gland on order of the hypothalamus, via the corticotropin-releasing factor (CRF). ACTH is responsible for regulating the secretion of the glucocorticoids cortisol and corticosterone from the adrenal cortex. Without ACTH or the glucocorticoids, we are unable to survive when exposed to any of a variety of stresses. Selye (1956) defined stress as the state manifested by the specific syndrome (general adaptation syndrome) which consists of all the nonspecifically induced changes within a biological system. Ganong (1963) has said that, "stress is certainly one of the most grandly imprecise terms in the lexicon of science. However, like sin, which also means many things to many people, it is probably here to stay, because it is a short, emotionally charged word for something that otherwise takes many words to say." Ganong has narrowed the definition of stress to include any of the multitude of stimuli that results in the liberation of increased quantities of ACTH.

Under normal conditions, glucocorticoid secretion is regulated by a feedback mechanism that slows ACTH secretion when the level of circulating glucocorticoids is high. However, when the brain interprets situations or stimuli as threatening, the hypothalamus calls on the anterior pituitary to secrete ACTH in an effort to counter the potential stress. Let us consider the effects of the stress response and then consider some ways in which exercise may act as a stressor.

THE ADRENAL CORTEX

The glucocorticoids of the adrenal cortex influence the metabolism of protein, carbohydrate, and fat, although their precise actions are incompletely understood. The glucocorticoids also exert a permissive action that enhances the effect of the catecholamines, epinephrine and norepinephrine, the hormones of the adrenal medulla. The importance of the glucocorticoids to the stress response may be related to that permissive action. They may be needed to maintain vascular responses to the catecholamines and to assist the catecholamines in the mobilization of free fatty acids (FFA) as an energy source. However, the essential nature of the glucocorticoids in the resistance to stress remains for the most part unknown. (Ganong, 1971.)

Exercise as a Stressor

Stress, tension, and aggressive personality patterns have been linked with ulcers, hypertension, heart disease, and various other ills that plague modern men. Numerous sources label exercise as a stressor. Paradoxically, exercise has received considerable attention for several possible cardioprotective mechanisms (see Chapter 11). Animal studies often conclude that exercise is a stressor. Frenkl and Caslay (1962) regularly forced rats to swim to exhaustion in a study of the effect of muscular activity on adrenocortical function. They concluded that the *exercise* produced a rise in adrenocortical synthesis and secretion. It is, of course, highly possible that the fear of drowning might also be considered a bit stressful. Other studies have forced rats to run on a treadmill. The rats received an electric shock when they drifted to the back of the treadmill. These studies also labeled exercise as the stressor.

On the other hand, some authors have been able to exercise animals without the threat of death or punishment. Miller and Mason (1964) enticed monkeys to climb in order to receive their food. They related the physical work done to the adrenocortical activity, and concluded that the psychological reaction to the task may have been more important than the physical activity. Suzuki (1967) conducted an elegant experiment to determine the effect of muscular exercise on adrenocortical secretion in dogs. The dogs, trained to run along with a bike-riding attendant, were then tested at varying distances and speeds to induce light to exhaustive exercise conditions. Analysis of adrenal venous blood revealed that only the exhaustive exercise resulted in marked increases in adrenocortical secretion. The authors concluded that the cortical secretion was related to exhaustion, but not to exercise, its duration or intensity.

Until quite recently most studies dealing with exercise and adrenocortical function in humans used indirect measures of cortical secretion. Bloodletting was avoided in order to eliminate the possible stress involved since earlier techniques required rather large blood samples. The corticosteroids secreted by the adrenal cortex appear in the blood and travel in a protein-bound form. Metabolic by-products are excreted in the urine, and this urinary excretion has been used to infer adrenocortical activity. Hill and his associates (1956) measured the urinary excretion of 17-hydroxycorticosteroids (17-OH-CS) of college oarsmen following practices, time trials, and competitive races. Practice measures were not elevated above control values, but both the time trial and competition values were elevated. The nonexercising coxswain had elevated values in one race but not in another. It should be noted that the demanding physical activity associated with the practice session did not provoke a significant rise in adrenocortical activity. Connell, Cooper, and Redfearn (1958) studied the effect of various physical and mental situations on 17-ketogenic steroid (17-KGS) excretion in fit young men. They found that physical exercise performed in the absence of emotional stimuli did not provoke an increase in 17-KGS excretion.

Steadman and Sharkey (1969) trained young male subjects on the treadmill, and measured their 17-KGS levels to determine the effects of exercise and training on adrenocortical activity. Early exposure to the strenuous treadmill test prompted a marked increase in 17-KGS excretion. As training progressed, the magnitude of the stress response diminished and eventually returned to the control level. This return to control levels occurred in spite of a steady increase in the daily work time on the treadmill. We felt that our large noisy treadmill and the subjects' uncertainty regarding the difficulty of the test contributed to the early elevation in 17-KGS levels.

Later, we used different subjects and duplicated the same experimental design, but this time we included measurement of whole blood-clotting time following the treadmill training sessions (Whiddon, Sharkey, and Steadman, 1969). The 17-KGS response was similar to that mentioned above: The clotting time decreased following the earlier exposures, but returned toward control levels as the stress response diminished. Hence, the exercise by itself did not influence the stress response nor the whole blood-clotting time. The clotting time seemed to be inversely related to the rise in 17-KGS.

Since urinary excretion of corticosteroid metabolites does not represent total cortical production, research using blood samples from humans is necessary to confirm the suspicion that exercise need not be considered a stressor. Hartley, Mason, Hogan, and others (1972) observed the effects of exercisc, exhaustion, and training on human cortisol valves. Light to moderate exercise had little effect, while exhaustive cffort caused a marked increase. Training did not alter the cortisol response to exhaustive effort.

Studies on dogs, monkeys, and men suggest that exhaustive, threatening, or competitive exercise may indeed stimulate increased adrenocortical activity. The available data does not indicate any exhaustion of adrenocortical hormones in even the most strenuous situations (Simonson, 1971). Furthermore, the adrenal hypertrophy often noted in animal studies does not seem likely in human subjects who engage in physical activity on a voluntary basis.

Stress in sport The physiological response to the stress of competitive exercise seems to be highly beneficial to performance (Fig. 8.1). The administration of adrenocortical hormone seems to prolong performance of exhaustive work, probably by mobilizing products that tend to spare the muscle glycogen. The permissive action of the glucocorticoids could also prove beneficial to the vascular responses necessary for vigorous activity. Thus, few of us would be willing to use the term *stress* in a derogatory fashion, especially among healthy, active individuals familiar with exercise and the stress of competition.

However, since stress may also provoke changes in clotting time and heart function (see Chapter 11), care should be taken in the prescription of exercise for less active adults, especially those in or approaching the coronary-prone years. Emotional

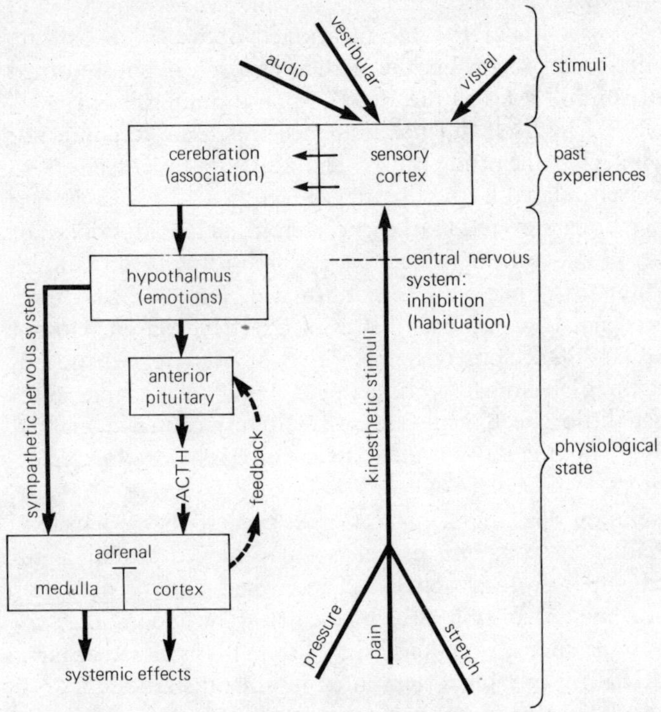

Figure 8.1. Stress and exercise. Adrenocortical hormones are secreted in response to ACTH stimulation. The catecholamines of the adrenal medulla are also under the control of the hypothalamus. Threatening stimuli activate the stress response. Our perception of the stimuli may be altered through experience with the event (e.g., exercise) or through a change in the physiological state (e.g., improved fitness). Repetitive exercise such as training may alter sensory input via CNS inhibition of incoming stimuli (habituation).

stress often occurs in the absence of exercise in our fast-paced, competitive society. Some of the elevated corticosteroid secretion that occurs with a psychological stress, such as an important examination, may be harmful since there is no physical activity to release the loaded stress mechanism (Connell, Cooper, and Redfearn, 1958). These factors suggest the potential value of, as well as the possible danger of, physical activity in the adult years. (They will be considered in greater detail in Chapters 11 and 12.)

THE ADRENAL MEDULLA

The adrenal medulla, although located within the adrenal cortex, is functionally and morphologically different from the cortex. The hormones of the medulla are epinephrine and norepinephrine, the catecholamines. In man, four-fifths of the secretion is epinephrine, but norepinephrine is also secreted from the nerve endings of the sympathetic portion of the autonomic nervous system. Adrenal medullary secretion may result from localized sympathetic activity, as during exercise, or via sympathetic nervous stimulation emanating from the hypothalamus. Thus, it is not surprising that situations provoking increased ACTH also prompt a rise in catecholamine secretion. In emergency situations the catecholamines prepare us for fight or flight. The heart rate and blood pressure increase, blood-clotting time is reduced, the blood vessels of the skin constrict, and glucose and free fatty acids are mobilized. As you remember, the adrenocortical hormones seem to be important for effective catecholamine function.

Simonson (1971) summarizes evidence indicating possible depletion of catecholamines in prolonged strenuous exercise. He suggests that such a depletion could be an important factor contributing to the eventual depletion of muscle glycogen during prolonged strenuous effort. The depletion of the catecholamines would lead to a decline in fat mobilization and a fall in blood glucose as well (Fig. 8.2). Consequently, the muscles would have to rely on stored glycogen as a source of energy.

The ratio of epinephrine to norepinephrine secretion seems to change according to the situation. Norepinephrine secretion is increased by psychological stresses that are familiar, while epinephrine is elevated during unfamiliar situations. We might hypothesize that a rise in norepinephrine occurs during familiar but stressful athletic competition; indeed, studies have found a greater rise in norepinephrine during athletic competition.

Some have speculated that anger directed outward is associated with norepinephrine secretion while anger directed inward is related to epinephrine. Others have gone so far as to

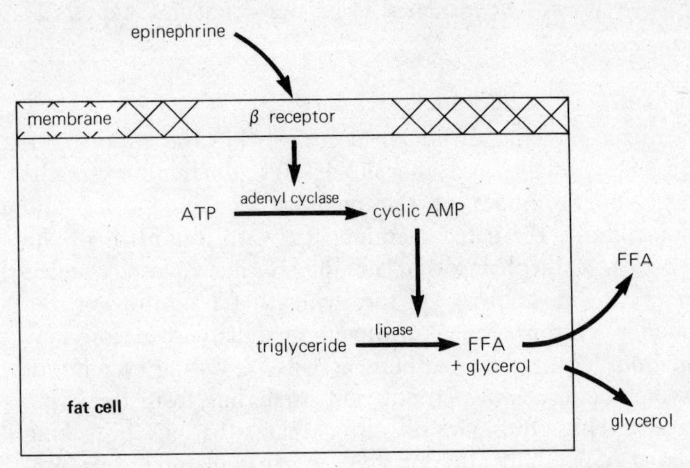

Figure 8.2. Mobilization of free fatty acids from adipose tissue. Epinephrine stimulates the cell membrane and activates the fat-splitting enzyme lipase, which then splits the stored fat molecule into three fatty acid molecules for transport to the working muscles. Growth hormone and the glucocorticoids also activate lipase while insulin inhibits its activation. Lactic acid may inhibit the influence of epinephrine and the mobilization of FFA.

suggest that a particular reaction may be learned via reinforcement of either the anger-out or anger-in behavior. These suggestions could have tremendous significance for the conduct of physical education and sport programs.

Does exercise in the absence of emotional or competitive stress provoke an increase in catecholamine secretion? Docktor and Sharkey (1971) studied the effects of exercise and training on the urinary excretion of catecholamine metabolites. Initial exposure to an unfamiliar and threatening treadmill test provoked a marked catecholamine response in untrained subjects. After several weeks of training, the postexercise response declined toward control values, in spite of a considerable increase in total exercise time on the treadmill. Frankenhauser et al. (1969) studied catecholamine excretion and subjective reactions to different work loads. They concluded that an increase in catecholamine excretion occurs during exercise when the effort grows so great that feelings of emotional stress, and not the actual work load, may determine the magnitude of catecholamine secretion from the adrenal medulla. Hartley, Mason, Hogan and others (1972) tested the catecholamine response to graded work loads of the bicycle ergometer. While the norepinephrine secretion increased with the work load (cardiovascular response), the epinephrine only became significantly elevated when the work load approximated the maximal oxygen intake, that is, when the exercise became unpleasant. Training reduced the norepinephrine response at all levels of effort and reduced the epinephrine levels at all work loads short of exhaustive effort.

Implications for Physical Activity

Friedman and Rosenman (1973) reported that a particular behavior pattern (type A) was found in association with coronary artery disease so frequently that it appeared possible that the behavior pattern itself may be involved in the etiology of the disorder. Workday catecholamine excretion was compared in 12 type-A men (aggressive, time conscious, competitive) and 11 type-B subjects (behavior opposite that of type-A men). Norepinephrine excretion was significantly elevated in the urine of the type-A individuals. While the authors were careful to avoid a cause-effect implication of behavior, stress, and coronary artery disease, they did ask some provocative questions regarding the relationship of type-A behavior and cardiovascular disease. (Byers et al., 1962.)

You will remember that norepinephrine secretion seems related to familiar psychological stresses, such as competition. The suggestion that a competitive way of life may be injurious to the health is not a new one, but experience shows that many of us are able to continue to enjoy the excitement it brings. However, it seems equally clear that some are not able to withstand the emotional and physiological assaults of excessive competition during the adult years. The implication of physical activity in both the etiology and prevention of stress-related diseases deserves careful consideration, since unfamiliar, highly competitive, or exhaustive physical activity may be regarded as potentially stressful, we must learn when and how such activities are tolerated. We must also learn more about the value of exercise in the reduction of psychological stress and tension. The uncritical endorsement of physical activity as a means toward better health must pass as did the era of leeches and bloodletting. It must be replaced by a reasoned utilization of the values of exercise coupled with a careful avoidance of its potential dangers. Such is the task confronting physical educators today.

ANDROGENS

Androgens are steroid sex hormones that are masculinizing in their action. The male testes secrete large amounts of the androgen testosterone and small amounts of feminizing estrogens. Androgens have *anabolic* or growth-increasing characteristics as well as *androgenic* effects that influence the appearance of masculine secondary sex characteristics. Hettinger (1961) reported that muscle "trainability" was enhanced when testosterone production was high or when it was administered as a supplement to older subjects. Muscle strength and morphology (cross section of fibers, amount of protein) were enhanced in dogs following the administration of the steroid. However, Fowler, Gardner, and Egstrom (1965) did not detect an improvement in athletic performance for young men using anabolic steroids. Johnson and O'Shea (1969) were able to demonstrate significant strength gains for their weight-training subjects. The experimental subjects who were administered an anabolic steroid also experienced a significant weight gain over the age- and weight-matched control subjects. The lack of a placebo group (i.e., those receiving fake medication) for psychological control makes it difficult to generalize from the results of this study. In a subsequent study utilizing a placebo group, Johnson et al. (1972) again found significant gains in muscular strength and size for the subjects in the anabolic steroid treatment group. In this and other recent anabolic steroid studies the treatment was successful when accompanied by a high-protein diet and severe muscular training stimuli. This interesting development will be considered further (see Chapter 10) when the influence of diet on human performance is discussed.

In spite of the recent evidence suggesting the effectiveness of anabolic steroid administration, many are concerned with the potential undesirable side effects associated with their misuse. Balance possible liver damage, prostatic hypertrophy, and testicular atrophy against the benefits to be gained (N.C.A.A., 1972). The greatest danger would seem to be during adolescence, the time of critical growth and development. The administration of exogenous anabolic steroids during these years could retard the normal sexual maturation and accelerate closure of the epiphyses, the growing portions of the long bones. We must wonder if a questionable improvement in athletic performance is worth possible sexual impotence or sterility, or stunting of long bone growth. Immature young men are less likely to receive a performance benefit and more likely to experience undesirable side effects. Possible uses and side effects have not been adequately explored among physically active older men.

OTHER HORMONES

Several other hormones deserve a brief mention because of their relationship with some aspect of physical activity. We have already mentioned the role of *insulin* in normal carbohydrate metabolism, noting that exercise allows the muscular utilization of glucose in the absence of insulin. *Glucagon*, another hormone of the pancreas, is secreted when the blood sugar is low. Together, insulin and glucagon serve to maintain the blood glucose level. Glucagon elevates the blood sugar via the breakdown of liver glycogen. At the same time it stimulates the secretion of the insulin necessary for the utilization of the glucose.

Parathyroid hormone is necessary for normal calcium metabolism. Calcium is essential for blood coagulation, cardiac and skeletal muscle contractions, and nerve function. Of course, calcium is also an important constituent of the bones. It is well established that bones respond to physical stresses by strengthening the stressed area. Prolonged periods in a weightless environment eliminate the gravitational pull and allow a loss of calcium and a weakening of the bones. Exercise seems to retard the demineralization that occurs during inactivity.

(Lamb, 1968.) However, little is known regarding the influence of exercise and training on the parathyroid hormone and calcium metabolism.

Aldosterone, a mineralcorticoid secreted by the adrenal cortex, is important for the reabsorption of sodium from the urine as well as from the sweat and other body fluids. Its secretion may result from ACTH stimulation, from high plasma potassium or low plasma sodium levels. Another cause of aldosterone secretion is *renin*, a substance secreted by cells in the kidneys. A decrease in renal arterial mean pressure increases the secretion of renin which, in turn, activates an increase in aldosterone release. The aldosterone promotes sodium and water retention which tends to raise the renal arterial pressure. It is easy to see how these factors may operate to conserve sodium during work in a heated environment. Since aldosterone secretion occurs with high levels of ACTH, it appears that its release is yet another consequence of the stress reaction.

PART III PHYSICAL ACTIVITY: PERFORMANCE AND HEALTH

We have reached the third and final part of our monograph dealing with physiology and physical activity. This section will attempt to "put it all together," to deal with *practical applications* concerning the physiology of training and limits of human performance. It will also consider the relationship of physical activity to cardiovascular health, and the importance of physical activity and fitness in the control of body weight. Chapter 12 will offer a practical guide to the prescription of exercise for physiological fitness and weight control.

Researchers are typically cautious about making practical suggestions based on their limited laboratory findings because generalizing beyond one's data is the cardinal sin of research. Attempts to apply laboratory findings in the real world always run this risk. However, we have attempted to base our generalizations on the *best* available evidence. So we gladly accept the risks involved as we attempt to communicate the practical outcomes of a complicated discipline to future physical educators, coaches, and others interested in the fascinating world of physiology and physical activity.

CHAPTER 9 THE PHYSIOLOGY OF TRAINING

Training for athletic competition or for personal improvement in a favored sporting event can be based on sound physiological principles. The training program should take into account the demands the activity places on the physiological systems and reflect the capabilities and deficiencies of the individual athlete. The program can be individualized to provide the best possible activities and experiences to assist the athlete to meet his or her competitive goals. In the development of the individualized program you should consider various training principles (e.g., overload, specificity) as well as the influence of boredom and fatigue. Finally, the daily training plan should take into account principles of motor learning such as progression and distributed practice. So you can see that the development of a training program is a complicated task requiring careful attention and study on the part of the coach. Let us take a closer look at the steps involved in the preparation of a modern, physiologically sound training program.

PRINCIPLES OF TRAINING

As a coach, one of your initial tasks is to analyze the sport or event to determine the particular demands it places on the athlete. You should first determine the major sources of energy required to perform the activity (Fig. 9.1), and then list other specific requirements that contribute substantially to success.

SOURCES OF ENERGY	*SPECIFIC REQUIREMENTS*
Aerobic	Skills (and mechanical efficiency)
Anaerobic	Muscular strength
phosphagen	speed
lactate	power
	Muscular Endurance
	Flexibility

The program should utilize the *overload principle* to develop the energy source required in the sport and to improve the specific factors involved. As an example, let us consider the sources of energy for a performer in gymnastics. Since most routines are less than 2 minutes long, the sport would not require a large aerobic capacity. As a matter of fact, gymnasts do not score much above active but untrained individuals in aerobic capacity. However, gymnasts do need anaerobic energy to complete skilled routines. They require specific muscular strength and endurance to practice and perform their specialties. But like most athletes, they will be better able to withstand the rigors of training if they improve their oxygen transport and utilization systems or aerobic capacity.

Your decision regarding a program for a distance runner would be far less complicated than that for a gymnast, or so it would seem. The distance runner requires a well-developed

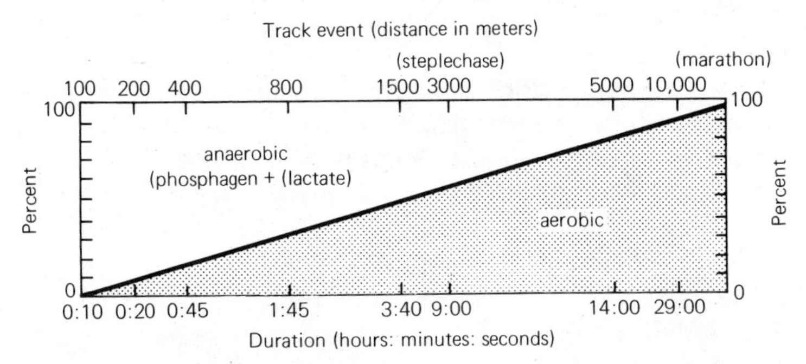

Figure 9.1. Aerobic and anaerobic energy sources in relation to distance and duration of event. Since exercise intensity must decline as the distance or duration of an event increases, we see a decline in anaerobic energy and a rise in the percentage of the aerobic contribution to the energy demands of the event. Events in other sports (e.g., swimming) may be analyzed in terms of the duration of the event.

(*Source:* Adapted from Wilt, 1968.)

aerobic capacity. Your training program would emphasize the overload of the oxygen transport and utilization systems with long-distance and interval training techniques. Your real problem would be to help the athlete maintain interest in the program. The coach should be able to assess the athlete's capabilities for the activity and offer advice regarding his strengths and weaknesses. A young man with a low aerobic capacity lacks the natural endowment to become a successful distance athlete. You as the coach should attempt to determine his capabilities and counsel him accordingly. On the other hand, should he desire to persist in endurance events it is your job to provide him with the best possible advice regarding his training. As an alternative he might be interested in some activity that combines endurance and skill, such as cross-country skiing or speed skating.

Training activities should be designed according to the principle of *specificity*. You should remember that in training, as in life, you get what you work for. This means that strength, endurance, or glycogen supercompensation activities should be accomplished in the muscle groups or motor units to be used in competition. You should also attempt to utilize the movement patterns to be used in the sport. This also suggests that a portion of your practice should be accomplished at the pace to be used in the event. Training at race-pace ensures the overload of the specific motor units to be used in the race, and it ensures the overload of the specfic energy sources utilized at that pace.

The training program should allow adequate *recovery*, either from one training task to another or from day to day. If you do not allow adequate recovery between interval runs, the athletes will be unable to maintain the pace or form desired or to complete the total workout. Should the daily practices be so demanding as to extend recovery beyond a 24-hour period, the physiological compensations will be inadequate and improvement will be negligible. Most effective training programs utilize a hard-easy (alternate day) principle that allows adequate recovery and progress from day to day. The definition of hard and easy is a relative one that depends on the age, fitness, and experience of the athlete. What is hard for a junior-high runner may be more than easy for a high-school distance ace.

The *skills* and mechanical efficiency required in a sport should be developed along with the sources of energy. Training programs should not ignore the practice of skills, and emphasis should be placed on the repetition of skills at the speed to be used in competition. Many skills are rate-dependent and accuracy developed at a slow speed may be lost when the movement is accelerated. Many successful coaches utilize this concept to gain a competitive advantage. When an opponent is forced to rush a jump shot in basketball or a backhand in tennis his performance will suffer. During practice, drill the specific skills at the pace to be used in competition. Of course, the training program should emphasize the *development* of necessary skills and should incorporate the principle of *progression* throughout the season. You should not start a new diver on difficult dives; he must first learn a good approach, take-off, and entry before he can master the two-and-one-half in the pike position.

In this age of specialization it is not surprising that successful competition at advanced levels now requires *year-round* attention to training. In a way it is regrettable to see 10-year-olds specializing in swimming, to the exclusion of other athletic or even social activities. It is sad to see the early overorganization of sports such as Little League baseball, football, or swimming. It is regrettable that coaches or even dance teachers forbid their pupils to engage in other activities, such as recreational skiing, because they may somehow interfere with the sport (or dance) performance. Success in sport has become an end in itself, and the dedication required for such success includes almost continuous attention to training and practice.

But from a physiological standpoint, long-duration training makes good sense. It takes many months to build the functional and structural bases for arduous endurance performances. Bones and ligaments as well as muscles must adjust to the rigors of contact sports or even endurance contests. Distance runners remain in some form of training throughout the year (Fig. 9.2). Long before their competitive season they begin the slow, steady development of the aerobic capacity. It is not unusual to find serious runners returning to training in July for a cross-country season that begins in October. During this period, the intensity of training is gradually increased. Anaerobic and speed work do not begin until a solid aerobic foundation has been established. While these volume workouts (often over 100 miles per week) may not produce vivid improvements in the aerobic

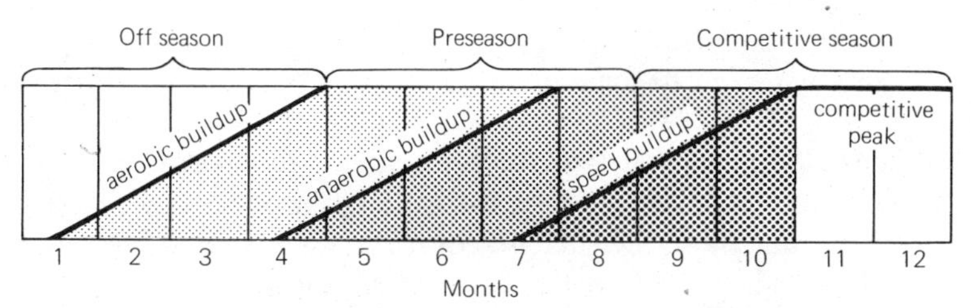

Figure 9.2. Annual training scheme. Begin aerobic training following a period of active rest and freedom from formal training. Introduce anaerobic work gradually after a solid aerobic foundation has been established. Begin appropriate speed work several weeks before competition. Skills should be learned early and refined during preseason training sessions. Strength gained in off-season weight programs may be lost during the pretraining and competitive periods unless appropriate supplementation is included.

capacity of already highly trained athletes, they do allow the runner to utilize a greater percentage of his maximal oxygen intake during prolonged effort. (Astrand and Rodahl, 1970.) It is this ability to continue to work at a high percentage of maximal effort that allows a well-trained and highly motivated athlete to disregard the discomfort of the task and hold on to defeat the opponent.

Similar principles apply to other sports. College football and basketball players now feel the need to practice or train throughout the year. Such attention to the development of energy sources, strengths, and skills will undoubtedly improve the quality of sport performances. However, such specialization may be wasted on all except the few who go on to careers in professional sport. In fact, certain aspects of specialization in sport raise questions regarding the coach's responsibility as a physical or health educator. The bulking-up of football players or weight men in track certainly seems to lead to success in athletics. But we must wonder what the future holds for the health of these magnificent athletes. Will they be able to maintain the muscle tissue they worked so hard to develop? Will they be able to curb the appetite they developed during their years of training? Or will they fall into the statistical trap that suggests that muscular but overweight men (endomesomorphs) are more frequent victims of cardiovascular disease? (Spain, Nathan, and Gellis, 1963.) It may be that coaches will someday accept postseason weight-loss responsibilities as readily as they now promote preseason body-building programs.

TRAINING METHODS

The methods used to prescribe training activities for athletes differ from those used to prescribe exercise for adult men and women. They differ in terminology as well as in purpose (see Chapter 12). Athletes train to reach specific objectives in sport while adults exercise to attain certain health and fitness benefits, and perhaps to aid their performance in a favored pastime. Both prescriptions are based on the same factors: *intensity, duration,* and *frequency of training.*

Wilt (1968) has classified a variety of track training methods according to their contribution to aerobic or anaerobic capabilities or speed. Listed in order of their suggested contribution to the development of aerobic capacity they include:

Aerobic
1. continuous slow running
2. continuous fast running
3. interval sprinting (alternate sprint-jog for long distance)
4. slow-interval training

Anaerobic
5. repetition running (long distances followed by complete recovery)
6. speed play (fartlek)
7. fast-interval training

Speed
8. repetition of sprints

Wilt suggests that methods 4–7 contribute significantly to anaerobic capabilities, and that the repetition of sprints and acceleration sprinting both contribute to the development of speed. Remember to develop the aerobic foundation (items 1–3) before attempting demanding anaerobic training activities. Long-slow distance training (LSD) has emerged as the most popular method for the development of the aerobic foundation.

Many training methods can be considered as *interval training* techniques, where the athlete runs a given distance at a given pace, rests and then repeats the run-rest program until the prescription is completed (e.g., 440 yards run six times at 63 seconds). The prescriptions differ in terms of the rate and distance of the work intervals, the length of the rest intervals, and the number of repetitions of each work interval. The length of recovery periods or rest intervals may be individually determined by checking the pulse rate (at the wrist or carotid artery), and resuming exercise when the pulse has returned to about 120 beats per minute. Exercise intensity can be gauged from immediate postexercise pulse counts. Begin immediately, and take a 10- or 15-second count and multiply by six (or four) to get the rate in beats per minute (bpm). Most fast running will elicit heart rates at or near 180 bpm. Continuous long periods of slow distance running will be accomplished at a lower heart rate. Obviously, as the rate or intensity increases the distance or duration must decrease. Active rest or recovery intervals must be extended after high-intensity or long-duration efforts. Walking or slow jogging during the rest interval may enhance oxygen transport to the muscles and the removal of lactate.

Physiology of Interval Training

The apparent value of interval training is that it allows the athlete to train at a higher *intensity* for a longer *duration*. Thus, the total work load in a given training session is far greater than could be accomplished by continuous work. Continuous high-intensity effort will result in the rapid accumulation of lactic acid and subsequent fatigue. Intermittent periods of work and rest allow for a partial recovery of the phosphagen and lactate energy systems between bouts of exercise. Thus, an athlete can run 440 yards (with several minutes' recovery time) ten times at a pace that would exhaust him after several continuous laps.

The rate and distance of the training interval depends on the event or sport in question and the sources of energy involved (see Fig. 9.1). Events lasting less than 2 minutes require speed and anaerobic capabilities; those lasting from 2–6 minutes require both aerobic and anaerobic energy sources, and those lasting beyond 6 minutes utilize an increasing proportion of aerobic energy sources. The energy for the marathon run comes almost entirely from aerobic energy sources (97.5 percent; Wilt, 1968).

Individual training prescriptions must also be based on the capabilities of the athlete. Running speed can easily be evaluated in a short dash. Anaerobic power (phosphagen) can be predicted from an all-out stair run (Fig. 9.3). Total anaerobic capabilities can be inferred from the athlete's ability to maintain speed throughout a 330–440-yard dash, or in any all-out effort lasting 40–50 seconds. Aerobic capacity can be predicted using the average running speed per minute during a 15-minute run (Fig. 9.4).

Football players utilize anaerobic power in short, explosive bursts of energy. The interval between plays allows a partial recovery of the phosphagen system and this recovery is aided by an efficient oxygen transport (aerobic) system. Basketball

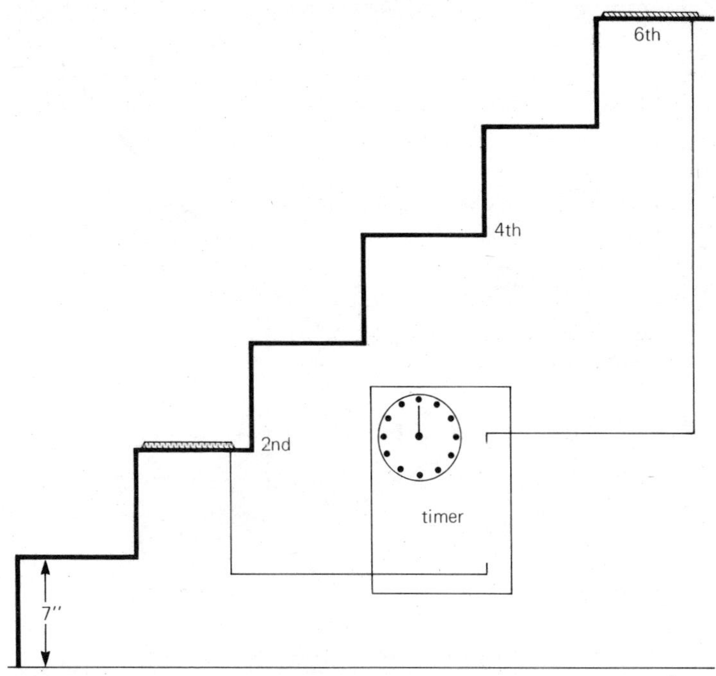

Figure 9.3. Anaerobic (athletic) power test. Maximal power (P_A) may be evaluated via a procedure suggested by Margaria et al. (1966). Following a brief approach, the subject runs up a flight of stairs, two at a time. The time required to cover the vertical distance between the 2nd and 6th step is used to calculate power.

$$P = \frac{F \times D}{t} = \frac{WT \times D}{t}$$

$$P_A = \frac{165 \text{ pounds} \times 2.33 \text{ feet}}{0.4 \text{ second}}$$

$$P_A = 960 \text{ foot pounds per second}$$

players require both aerobic and anaerobic capabilities. Their sport alternates the submaximal effort of ball control with the anaerobic burst of the fast break. Efficient oxygen transport and utilization allows better recovery during the submaximal periods, foul shots, and time-outs. Wrestlers require muscular strength and considerable muscular endurance (local aerobic ability).

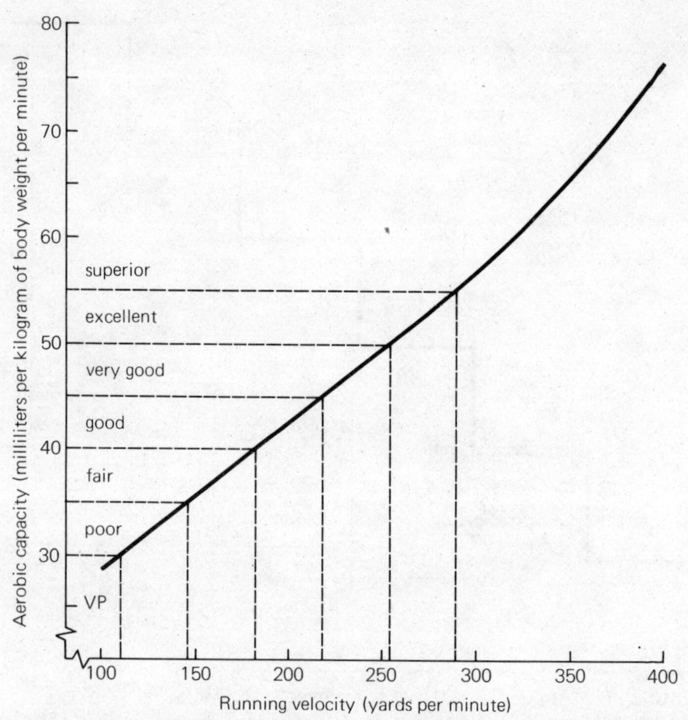

Figure 9.4. Predicting the aerobic capacity. After a light warm-up, rest and then run as far as you can in fifteen minutes. Divide the total distance by 15 to get your speed in yards per minute. Then consult the chart to determine your predicted aerobic capacity in milliliters of oxygen per kilogram of body weight per minute. For metric conversion, meters = yards × 1.094.

(*Sources of data:* Balke, 1963; Daniels, 1972.)

They, too, benefit from a well-developed aerobic capacity. Swimmers, like trackmen, require energy sources related to the duration of the particular event. Sprinters operate on anaerobic energy, while distance athletes utilize aerobic sources to a great extent. The repetition of submaximal contractions during swimming requires considerable muscular endurance. Cross-country skiers require a well-developed aerobic capacity with the anaerobic ability to tackle the uphill stretches. Alpine skiers function anaerobically in events that seldom exceed 2 minutes. They require considerable muscular strength in the legs to control the tremendous forces encountered in high-speed turns. *All athletes* are better able to recover and to tolerate long practice sessions when they develop the cardiorespiratory systems involved in oxygen intake, transport, and utilization (aerobic capacity).

TRAINING SYSTEMS

Some coaches advocate the mixing of aerobic and anaerobic training, while others are certain they should be separated. Some believe in a certain pattern of interval runs. Most successful coaches have developed some personal preferences based on their experiences over the years. However, neither the empirical observations of the successful coach nor the impersonal conclusions of the research literature point to a single effective system of training. As Wilt (1968) has written, "The problems of how far, how fast, and how often to run in training for optimum competitive results still remain largely unsolved."

Thus, the responsible coach cannot rely on others when it comes to the development of individualized training programs. You cannot avoid analyzing the demands of the sport and the capabilities of your athletes. Levels of fitness differ, as do age and sex. Since most track or swimming books are written by successful coaches of Olympic-caliber athletes, the training suggestions must be scaled to fit your young athletes. Frequent reference to the postexercise and recovery pulse rates will provide information regarding the severity of the program and its suitability for your athletes. Let us now consider how age, sex, and other factors influence the effects of training and the limits of human performance.

LIMITS AND IMPLICATIONS OF TRAINING

In this section, we shall consider the implications of several important but uncontrollable factors on human performance: heredity, race, age, sex and body type, all of which influence human performance capacity and impose limitations beyond the influence of training.

Racial Differences

We shall limit our discussion to some characteristics of the black population that may be related to their immense success in some forms of sport. We will avoid any effort to define race in a genetic sense, dealing instead with some of the theories brought forward to explain the relative success of the black athlete. How successful are black athletes? In spite of the low percentage of blacks in the total population of the United States, professional all-star teams in football and basketball included 44 and 63 percent blacks in 1970. (Kane, 1971.) While few observers were surprised to see black athletes dominate the sprints in the track and field portion of the 1968 Olympics, prejudices were forced to crumble as black athletes won all the distance events, including the marathon.

Black males are leaner than their white counterparts in the U.S. Army. Similar findings exist for adolescent boys and girls. Are these differences genetic or are they due to socioeconomic factors? Do black children score higher on fitness tests because of some inherited physiological factors, because they are leaner, or because blacks have a higher infant mortality rate that allows only the stronger ones to survive? Do black athletes in the United States stand out because of their longer and leaner extremities (Tanner, 1964), or because only the strongest and fittest of their ancestors were able to survive the passage to slavery and the arduous work it involved?

Environmental factors may have operated to provide both advantages and disadvantages to the black in sport. The relative leanness, possibly resulting from a warmer climate of origin, allows a greater power to body weight ratio. However, that leanness may be a disadvantage in swimming, or in winter sports such as skiing. Blacks are less buoyant and must, therefore, utilize a larger percentage of their energy expenditure to remain afloat. In cold climates their extremities are cooled more readily and their discomfort is more acute.

Anthropologist E. E. Hunt has noted that the black man's heel doesn't protrude as much as the Caucasian's, and in combination with a longer extremity he gains an advantage in leverage for jumping. (Kane, 1971.) While some might consider the success of the black athlete in basketball as proof of that advantage, others would note that several of the world's best high jumpers are white. We must agree with Malina (1972) who notes that there is more to human performance than "sheer muscle and bone. It is the individual who eventually must perform and a whole complex of human factors are involved."

Hereditary Limitations

Can anyone train long enough and hard enough to become a champion distance runner, or does breeding set limits on the factors involved in the aerobic capacity? We are all aware of great father-son or brother-sister combinations in sport that seem to indicate the genetic advantage to be gained in "picking your parents" carefully. Horse breeders and trainers have long recognized the importance of a good blood line in a horse race, and there is little doubt that selective breeding could eventually lead to improvements in some sporting events. However, most of us shudder to think that sport would ever become *that* important to the human race. Our question, then, is how important *is* heredity and how much can environment (in this case training) influence trainability and the limits of human performance?

Let us limit our discussion to the heritability versus the trainability of one important aspect of athletic performance, the maximal aerobic capacity. Various studies have indicated a trainable limit of 20–25 percent. Few researchers would expect even the least fit individuals to surpass their aerobic capacity by more than 30 percent. Thus, an unfit, inactive individual may be able to improve his aerobic capacity from 40 ml per kilogram of body weight per minute to 48 or 50 ml following an extended period of training.

Klissouras (1971) studied differences in aerobic capacity among 25 pairs of male twins aged 7–13 years. Young twins were chosen to minimize the effects of variable environments (e.g., diet, training) on aerobic capacity. The author concluded that variability in maximal aerobic power was 93.4 percent genetically determined. Intrapair differences were greater among fraternal (dizygous) twins than among the identical (monozygous) twins. The largest differences between pairs of identical twins ($50.34 - 46.84 = 3.5$ ml per kilogram of body weight per minute) was smaller than the differences typically measured between the ten pairs of fraternal twins. Thus, it seems that heredity plays an important role in determining the capacity for maximal human performances involving the maximal aerobic capacity. Klissouras also noted that variability in anaerobic capacity and maximal heart rate was 81.4 and 85.9 percent genetically determined.

Clearly, there are enormous variations in genetic endowment for athletic performance. Many coaches establish youth programs that allow the endowed youngsters to be identified. Some countries identify such talent and then make a systematic effort to develop it for nationalistic purposes. High-school, college, and professional recruitment of athletes represent efforts to assemble the gifted for athletic competition. These attempts are undoubtedly beneficial for the quality of the sport and, perhaps, because they direct the less gifted to other more suitable areas of endeavor. However, they may also serve to exclude interested but less gifted youngsters from the advantages available in the world of sport.

Sex Limitations

Victorian notions to the contrary, women seem quite able to bear the burdens of athletic training and competition. Athletic mothers seem to experience shortened labor and fewer cesarean deliveries. (Thomas, 1969.) Aside from occasional cramps during the period, there is no evidence to support a limitation of training or competition during any phase of the menstrual cycle. On the contrary, women seem to respond well to training, and sex differences in performance may be attributed to hormonal and secondary sex characteristics, differences in body size, or sociocultural factors such as frequency or opportunity of participation.

Muscular strength seems to increase more in young boys as they enter adolescence. This increase in trainability is probably due to the increase in androgen production by the testes. Fat deposition—another factor related to the sex hormones—also differs, and girls typically carry a greater percentage of their weight as adipose tissue. Hence, the already smaller girl has less strength per unit of body weight. The greater amount of adipose tissue becomes an advantage in swimming, especially distance swimming, where buoyancy and a reduced heat loss become important.

Before puberty, boys and girls do not differ in aerobic capacity, but from that point on women average about three-fourths of the men's capacity. (Astrand and Rodahl, 1970.) Since men have more hemoglobin (16 gm versus 14 gm per 100 ml) we might expect their oxygen-transport system to have a greater capacity. In fact, hemoglobin concentration and maximal oxygen intake are significantly related in women (Haymes, Harris, Beldon et al., 1972). However, recent animal studies question the importance of the hemoglobin difference. When hemoglobin and oxygen-carrying capacity were reduced in dogs (from 15.7 → 11.1 gm per 100 ml), oxygen delivery was maintained by an increase in the flow of the less viscous blood. (Horstman and Gleser, 1973.)

While it would not seem necessary to encourage athletic competition between the sexes, there seems to be *no* justification for the limitation of athletic opportunities for women. Women can and *do* compete favorably with men in many sports. Improved opportunities for practice, good coaching, and competition may well enhance women's future image in sports.

Age Limitations

Aerobic capacity (measured in liters per minute) and strength increase into the late teens or early twenties and then decline slowly with the years. It has been customary to suggest that peak performances in given athletic events are attained at certain ages. However, even that longstanding generalization deserves a closer look. Distance runners were said to mature in their thirties, and the data from the 1952 Olympics support that claim. But the marathon runners in the 1960 Olympics on the average were three years younger than those in the 1952 games (Jokl, 1964), and the winner of the 1972 marathon was only 24 years old. Young girls are achieving world class performances in various sports including swimming, speed skating, tennis, and track and field. Six-year-old boys and girls now enter into serious athletic training for swimming, and it is becoming increasingly difficult to achieve success without an early start.

Daniels and Oldridge (1971) observed changes in the aerobic capacity of young boys (10–15 years old) during a 22-month period of endurance training. Their aerobic capacity increased from 2.3 to 2.8 liters per minute. However, since their body size also increased there was no difference in the aerobic capacity expressed in milliliters per kilogram of body weight (59.5 ml per minute). The authors concluded that the improvement of performance in the mile run was caused by an apparent improvement in running efficiency, caused in part by an increase in linear growth. It would have been interesting to follow a control group during the training period to discern the effects of growth alone on performance in running, and to determine the benefits of prepubescent training on endurance performance.

There does not seem to be any physiological reason to eliminate early training experiences, so long as they are tailored for the age and experience of the participants. McManus and Lamb (1972) observed a dramatic and consistent retardation of muscle growth and body-weight gain in young guinea pigs exposed to strenuous endurance training, and they expressed concern regarding the excessive training of prepubertal children.

Early overspecialization may lead to disenchantment at an early age. Seventeen-year-old women and 20-year-old men have been known to retire from swimming while still capable of world class performances. Early training experiences will undoubtedly identify those with the natural endowment to succeed in a given sport and they may even help some athletes develop the skills and capacities necessary for success. But they may also stifle youthful exploration and lead to an early retirement during the years of peak physiological performance capacity.

Despite the decline in performance with age (Fig. 9.5), there is encouraging evidence that supports the value of training at all ages. The rate of decline in strength and aerobic capacity can certainly be slowed with training. The inevitable symptoms of aging are most pronounced during maximal aerobic or anaerobic efforts. In submaximal performances, where skill is an important factor, high-quality performances can be maintained well into the forties or beyond. Motor skills need not decline along with the physiological functions. Several successful golfers continue to win prize money in spite of their 50 years or more. Bobby Riggs continues to play competitive tennis in his fifties, and the amazing Clarence deMar continued to run the marathon long after his sixtieth birthday. And Larry Lewis continued to run six miles a day, every day, well beyond his one hundredth birthday! So, in spite of a decline in physiological

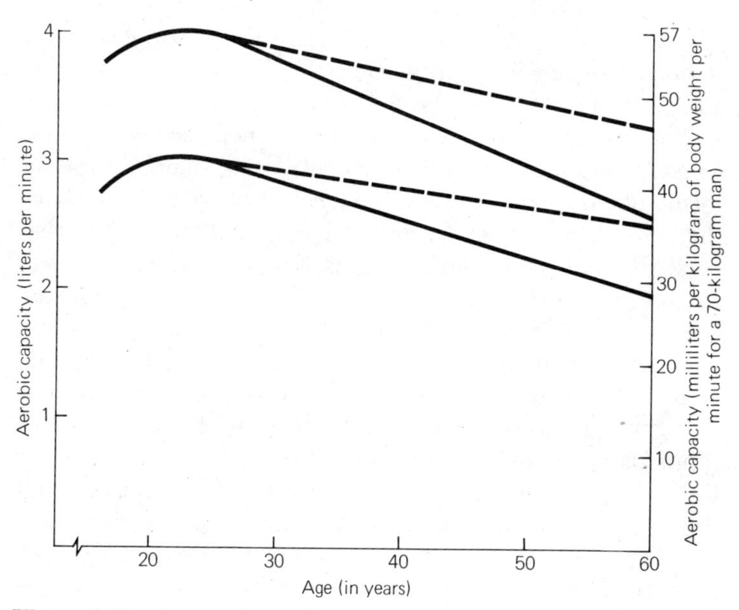

Figure 9.5. Age and aerobic capacity. The typical pattern of decline is illustrated with a *solid line*. The *broken line* suggests the benefit of continued activity. Note that the rate of decline is similar for fit and unfit individuals.

trainability with age, there is no reason to give up or to avoid athletic training because of age.

Limitations of Physique and Body Composition

When we speak of *physique* we mean to imply height and weight as well as the leanness, adiposity, or muscularity implied in the classic body-type designations (somatotypes) of ectomorph, endomorph, or mesomorph. In gross terms, height, weight, and body type relate to success in specific sports. Ectomorphic distance runners are shorter and weigh less than mesomorphic weight men in track. Generally speaking, tall basketball players are in greater demand. However, while many sprinters may be short and relatively muscular, some notable exceptions have been rather tall and slim. While physique alone may provide some insight for *counseling* young athletes' participation in specific sports, it does not tell you nearly enough to enable you to *predict* success in events where skill, training, and motivation are important components of the total performance.

Body composition techniques allow a more detailed analysis of body fat and the lean body weight. Using skinfold measurements of subcutaneous fat or underwater weighing procedures, the percentage of body fat and lean body weight can be calculated. If a 200-pound individual is found to have 10 percent fat, his lean body weight (LBW) will be 180 pounds (10 percent $\times$ 200 $=$ 20 pounds of fat). The high caloric consumption of strenuous exercise will result in a reduction in the percentage of body fat. While it is quite true that some athletes gravitate to a given sport because they are built for it, it is also true that sports participation is capable of exerting a profound effect on the body composition. Katch and Michael (1971) reported body composition findings based on skinfold measurements taken on 94 high-school wrestlers. Wrestlers in the light- to medium-weight categories averaged 4.9 percent fat while those in the heavier weights carried 11.7 percent of their weight as fat. Katch and Michael compared their values with data for normally active college men (12.5 percent), young male faculty members (16.5 percent), marathon runners (7.5 percent), high-school basketball and football players (7.9 percent), college basketball players (9.7 percent), college football athletes (15.3 percent), college soccer players (9.3 percent), and baseball players (14.2 percent) previously reported in the literature. Healthy normal college women range from 13.6 to 36.8 percent with an average of 25.7 percent (Wilmore and Behnke, 1970) while female endurance runners (15–26 years old) average 11.7 percent. (Brown and Wilmore, 1971). We will discuss the relationship of exercises to weight control in Chapter 11.

Record Performances

Throughout the years various authors have attempted to predict the future of world record performances, especially in those events that allow accurate quantification, such as track or swimming. However, such predictions are necessarily based on what is now known about performance, physiological capacities, and limitations. They do not take into account such new developments as the artificial track or the fiberglass vaulting pole. They are limited by previous knowledge concerning the limits of certain physiological functions (e.g., recent studies have demonstrated far greater maximal oxygen-intake values as well as higher cardiac output measures than were previously recorded or believed possible). These attempts at prediction are valuable because they force the integration of the available knowledge into a testable theory of maximal human performance. However, some of us will continue to hope that the uncertainty of sport will never be reduced to a computerized prediction based on physiological capacities and anatomic dimensions. It seems safe to predict that improved early nutrition, training techniques, and equipment, as well as an increase in the number of participants will continue to result in an assault on the records. However, changes in life style or altered competitive values could easily nullify that conservative estimate.

CHAPTER 10 FACTORS INFLUENCING HUMAN PERFORMANCE

Let us now consider some of the factors that have been suggested as aids to human performance. We will begin with a discussion of nutrition and then consider various drugs and other possible ergogenic aids.

NUTRITION

In earlier days when sportsmen came from the wealthy, well-fed class, the nutrition of athletes was not considered a significant problem. But today when athletes come from all socioeconomic levels, we are beginning to realize that adequate nutrition is essential if an athlete is to train and compete successfully. Recent evidence suggests that physiological as well as intellectual functions may be retarded if adequate nutrition is lacking during crucial periods of early growth and development. Let us consider the effects of under and overnutrition on performance as we attempt to define what constitutes good nutrition for an athlete.

Food Intake and Energy Expenditure

The increased energy expenditure of athletic training and competition requires an increase in the dietary caloric consumption if the body weight is to be maintained. Jokl (1964) has shown the detrimental effects of low caloric consumption on athletic achievement among the nations of the world. The winning of Olympic medals seems related to adequate nutrition. In daily training, the athlete will burn 300 calories or more above the usual daily expenditure. The marathon runner may consume an additional 2500 calories in training or competition. Wilderness backpackers and mountain climbers engage in prolonged submaximal activity that burns over 3000 calories above the amount consumed during a typical day at the desk (about 2600 calories). What proportion of carbohydrate, fat, and protein should be included in the athlete's diet? How much of each should he ingest to achieve optimal nutrition and performance benefits?

In normal nutrition the carbohydrate, fat, and protein proportions may average 50, 35, and 15 percent respectively. There is considerable debate at present regarding the health implications of fat in the diet. Some suggest the need to reduce the total amount of fat ingested while others argue for a reduction in saturated fats. Athletes seem to elect a diet that is lower in fat but higher in protein. It may be that they select protein because of the common misconception that protein serves as a source of energy in vigorous activity. On the contrary, we have already noted the fact that protein is employed as an energy source only during periods of starvation. More important than the quantity of protein or fat ingested is its quality, because certain amino acids as well as free fatty acids (FFA) cannot be synthesized in the body. The essential amino acids must be included in the diet (meat and animal products are a good source for these). Failure to supply one of the essential amino acids will hold up protein synthesis. The dependence on plant proteins in some countries accounts for the prevalence of protein-deficiency diseases.

Protein

We have discussed the use of carbohydrate and fat as sources of energy for muscular contraction and the value of glycogen supercompensation for prolonged endurance events. Now let us briefly consider the need for protein and the effects of protein supplementation on athletic performance. The amino acids of the protein we ingest are utilized to build cell walls, contractile tissue, hormones, enzymes, and a variety of other molecules. Studies have shown that adult protein needs do not increase markedly during physical activity. In fact, it appears that daily protein requirements may be met with as little as 1 gm per kilogram of body weight. Thus a 70-kg man (154 pounds) would require about 70 gm or 2.5 ounces of good-quality protein daily. Small wonder that the urine of the pampered athlete is a rich source of nitrogen!

When total protein needs are not met and when the essential amino acids are missing from the diet, physical activity will result in a loss of muscle mass. However, protein supplementation, beyond that which is needed to maintain nitrogen balance, has not proved beneficial to human performance. It seems logical to assume an increased need for protein during training programs that provoke contractile (strength) or enzymatic (endurance) protein synthesis. However, the evidence suggests that 1 gm of protein per kilogram of body weight provides the reserve to meet even those needs, unless the ill-advised use of anabolic steroids has altered the normal nitrogen balance. An increase to 2 gm per kilogram of body weight will provide a margin of safety for the most strenuous of training programs (Yoshimura, 1965), and will meet the growth and development needs of the youthful athlete. There is no detrimental effect of increased protein ingestion so long as the diet includes an appropriate balance of carbohydrate and fat. There does not seem to be any justification for the use of high-protein supplements in the nutrition of the athlete *on an already adequate diet* (Simonson, 1971).

Usually the *pregame* or precompetition meal need not include any special ratio of foodstuffs so long as it is easily digestible and readily cleared from the stomach. Some athletes get so excited they find it difficult to eat at all. Balanced liquid meals offer palatable and easily digested substitutes for solid food. Fat and protein could be reduced in the pregame meal if the competition is likely to involve an all-out endurance performance. The fat tends to slow the emptying of the stomach and protein metabolism lowers the *p*H. The higher ratio of carbohydrate will help to maximize the effects of glycogen supercompensation.

Vitamins and Minerals

Certain B vitamins serve as co-factors on enzymes involved in carbohydrate metabolism. Thus, it is not surprising to find that vitamin needs increase with physical activity. But since caloric intake is often increased to provide energy for the exercise, the vitamin needs should be met with the increase in an already sound diet. Various vitamin supplements have been tested to determine their effect on performance. Some few studies have reported improved performances following increased vitamin ingestion, but it is likely that the effect was due to the improvement of previously inadequate nutrition. Studies on vitamins A, B, C, D, and E suggest that no type of vitamin supplementation will improve any type of performance in an individual whose nutrition is adequate. (Simonson, 1971.)

Excessive vitamin supplementation (especially vitamins A and D) may carry some undesirable side effects. Some recent studies suggest that the actual minimum requirement for vitamin C may be less than the currently listed minimum daily requirement. However, many individuals are taking massive doses (3–4 gm) in hopes of avoiding the common cold. Most of the water-soluble vitamin C will appear in the urine. But while in the body, massive doses may encourage ulceration of the gastric wall, leaching of calcium from the bones, kidney complications and other problems more troublesome than the cold itself. While some have suggested the use of vitamin C to aid resistance to stress and recovery from physical activity, the studies in these areas are by no means conclusive.

The need for minerals and trace elements may increase with exercise, but these needs should be met with a commensurate increase in the normal diet. Some recent evidence indicates that female athletes may have low iron levels and should pay special attention to dietary iron intake. (Buskrik and Haymes, 1972.) Organic or health foods do not seem to provide added health or performance benefits. Thus, it appears that those on an already adequate diet will not benefit from nutritional supplementation, aside from the proved preperformance benefit of glycogen supercompensation. Undernourished athletes may benefit from an improved nutritional status, but it seems that their younger brothers and sisters may have far more critical nutritional needs.

Weight Gain and Weight Loss

Athletes are often interested in gaining weight in order to improve their chances in a sport such as football. This is sometimes a difficult task for extremely active young men. The usual attempt to force-feed carbohydrates (usually starches) proves rather disappointing since carbohydrate provides only 1860 calories per pound while fat provides 4200 calories per pound. Thus, when athletes ingest large amounts of fat-containing meat they acquire a larger portion of the calories they desire. Bourne (1968) suggests that it may be almost impossible for those engaged in heavy work to obtain sufficient calories via carbohydrate ingestion without discomfort. The nutritionally sound and palatable liquid meals provide another way for the athlete to add calories between his regular meals.

Weight loss can be achieved in a sensible program of caloric restriction combined with an increase in caloric expenditure via exercise. We shall discuss the interrelationships of exercise, fitness, and weight control in Chapter 11. At this point we should consider the implications of rapid and extreme weight loss as practiced in wrestling. There are two methods commonly used to "make weight" in wrestling. One employs sound physiological principles involving caloric reduction and an eventual loss of body fat. The other method is medically and physiologically unsound since it involves the loss of body fluids (dehydration) and a decline in both strength and endurance.

The most efficient way to lose weight is to stop eating. However, such an extreme practice should be undertaken only under close medical supervision. Furthermore, it is unreasonable to expect an athlete in training to go without nourishment. A program combining caloric restriction and an increase in energy expenditure will lead to the desirable weight loss without the loss of strength and endurance. Should an athlete with a preseason caloric intake and expenditure of 3000 calories restrict his intake to 2000 calories daily while he increases his exercise expenditure by 750, he will be forced to utilize his fat stores to provide the balance ($3750 - 2000 = 1750$ calories). About 3500 calories must be expended to lose each pound of fat, so that the athlete in our rather extreme example could lose a pound in 2 days ($2 \times 1750 = 3500$). Because the body needs some fat it would be physiologically unsound to reduce the body fat below 5 percent. The reduction to the desired level should be based on a knowledge of the subject's lean body weight and percentage of fat, and should be planned well in advance of the date of competition.

We have already noted that an exercising athlete may lose up to 3 liters of sweat per hour. Since each liter weighs 2.2 pounds the sweat loss would appear as a rapid and extreme weight loss. Dehydration weight loss is unsound for several reasons. If followed to extremes it may alter the electrolyte balance and result in convulsions or even death. Dehydration compromises circulatory efficiency and reduces endurance. It significantly lowers muscular strength. Coaches may rationalize the procedure by saying that the athlete is able to take fluids after the weigh-in and rehydrate prior to his competition. However, the rehydration effort may fall short of adequate fluid replacement. Thus, the coach could be asking his athlete to perform at less than maximal strength and endurance.

It seems likely that the wrestling world must soon adopt a physiologically sound and objective method for the determination of the minimal competitive weight. This method must be employed well in advance of the competitive season to allow the gradual loss of body fat over a sensible period of time. The method should be based on accurate measurements of the percentage of body fat and the lean body weight. Estimations of body composition appear to provide a logical solution to this perennial problem.

The Tcheng–Tipton formula (Tcheng and Tipton, 1972) for the prediction of a minimum wrestling body weight (MWW) is as follows:

$$\begin{aligned} \text{MWW} = {} & 2.05 \times \text{height (in.)} + 3.51 \times \text{chest depth (cm)} + \\ & 3.65 \times \text{chest width (cm)} + 8.02 \times \text{left ankle (cm)} \\ & + 1.96 \text{ bitrochanteric diameter (cm)} - 282.18 \end{aligned}$$

The formula is based on the assumption that weight loss should not continue below a minimum of 5 percent body fat (the minimum wrestling body weight). The method correlates well ($R = 0.923$) with accepted body composition methods. The procedure provides a reliable and objective means of determining a physiologically sensible minimum wrestling weight.

DRUGS

In our drug-oriented society, conditioned as we are to pain killers, tranquilizers, diet pills, depressants such as alcohol, and uppers such as amphetamines, it is not surprising to find professional, college, and even high-school athletes experimenting with drugs. Coaches, trainers, and team physicians accept the use of pain killers such as Novocain and antiinflammatory agents such as cortisone injections. Parents lament the widespread use of marijuana among the youth as they sip their predinner cocktails. When all of this is coupled with an intense desire to win in athletics, it is not surprising to find that the world of sport is deeply interested in the effects of various drugs on maximal human performance. Laboratory investigations may be inadequate tests for this purpose because they fail to simulate the complex emotional climates associated with competition, or because they deal with an isolated aspect of the total performance. With these limitations in mind, let us briefly survey some of the popular drugs, their physiological action, and possible effects on performance.

Stimulants

Central nervous system (CNS) stimulants or uppers, currently in use, range from potent amphetamine compounds or speed to those on the low end of the speed spectrum such as caffeine. The chemical structure and action of *amphetamine* is similar to that of epinephrine. Since the drug seems to reduce the sensation of fatigue it has been tested in a variety of endurance performances. However, conclusive evidence is lacking to support a definite and dependable performance benefit. The use of amphetamines to get "up" for competition is probably a mistake because the athlete may feel that he is performing at a higher level than he actually is. And because athletes react in a variable fashion to specific dosages, skill performances could actually suffer, and some might extend themselves beyond normal, safe limits. Finally, the potential for psychological dependence should override any of the limited and debatable benefits to be gained.

Caffeine In spite of somewhat similar effects on the CNS, caffeine is never recommended as an ergogenic aid. Some common sources include a cup of coffee with 150 mg of caffeine, a cup of tea with slightly less, and a glass of cola with about 50 mg. Excessive use leads to nervousness, irritability, and impairment in coordination, as well as to gastric distress and a loss of sleep. While caffeine is used medically to dilate coronary vessels it also acts as a diuretic (never a problem before a game!). The limited benefits of caffeine are far outweighed by the disadvantages when one is concerned with athletic performance.

Cocaine Cocaine is another powerful CNS stimulant that exaggerates the actions of the catecholamines. It seems to mask the sensations of fatigue and allow prolonged endurance efforts. However, it may also mask or override normal levels of inhibition and allow overstrain or injury. Cocaine is both habit-forming and illegal, and should not be considered as an aid to performance. None of the CNS stimulants mentioned seem worthy of further consideration in sport. Their benefits are not well documented, they affect different individuals in a variable fashion, they are often addictive, and they seem to be in conflict with the spirit of friendly competition.

Depressants

Alcohol, barbiturates (downers), or tranquilizers—seldom used to improve performance because of their detrimental effect on motor skills—are often used to "come down" from the high of competition. Small amounts of alcohol have also been found to relax the tense performer. However, since dosage is again dependent on factors such as tolerance or previous experience, it is difficult to predict the effect on performance. The need for downers to return from a high induced by uppers seems to be reason enough to avoid both extremes. There is not much evidence to support the use of depressants before or during competition. The fact that some may require assistance to relax following athletic competition may suggest the need to investigate the nature of the competition rather than the possible use of drugs.

Psychogenic Drugs

Pleasurable sensations, hallucinations, delusions, and bizarre behavior have been produced by a variety of drugs ranging from marijuana to LSD. While these drugs have not been suggested as possible aids to athletic performance, they have been implicated in the exodus of several talented and publicized athletes from the sport scene. LSD seems to exert its unpredictable effects via the inhibition of serotonin, an important inhibitory and tranquilizing transmitter substance in the brain. (Thompson, 1967.) This inhibition of inhibitions leads to hallucinations, and sometimes even results in recurring trips that have been linked to psychosis and suicide. It has been suggested that the drug orientation in society leads to experimentation and a search for new highs. The acceptance or even dependence of athletes on drugs may also lead to a climate for experimentation that leads to psychological or physiological addiction. The coach, athletic trainer, or team physician cannot ignore his contribution to the drug orientation in our culture.

OTHER FACTORS

Negatively ionized air, ultraviolet rays, and oxygen inhalation are other factors that have been said to influence human performance. The ionized air of the health spa has not been shown to produce any effect on performance. However, Hettinger (1961) has reported a beneficial effect from exposure to ultraviolet radiation, an effect that seems to be matched by the administration of vitamin D. This vitamin is produced in the skin by the action of sunlight. It is likely that the subjects were deficient in vitamin D, hence the effect of the ultraviolet rays on muscular performance. There is also the possibility that the tanning effects of the rays may have induced a feeling of well-being and a psychological effect on performance. The case for oxygen inhalation is a bit more complicated. Let us consider the various uses of oxygen as an aid to performance in sport.

Oxygen

The use of *oxygen* as a breathing aid goes back to 1774 when Lavosier inhaled the gas and found that his breathing became light and easy. In the 1936 Olympics the Japanese swimming team utilized oxygen before competition. Their tremendous success led to an increased use of oxygen as a performance aid, and to research dealing with the use of oxygen before, during, and after (recovery) performance. Oxygen inhalation received considerable publicity after some members of the U.S. ice hockey team were said to use it during their exciting Gold Medal performance in the 1960 Olympics. Oxygen tanks are frequently seen on the sidelines during professional and major college football games; and they are sometimes used during track and swimming meets, especially those conducted at higher altitudes.

Before performance Since valuable time is lost during each breath in a swimming race, it has been hypothesized that oxygen inhalation might aid breath holding and performance in that sport. Indeed, Karpovich (1934) demonstrated such an effect when the oxygen was inhaled *immediately* before performance. However, because oxygen cannot be stored in significant quantities and the blood is already 97 to 98 percent saturated while breathing atmospheric air, oxygen inhalation would not seem to be necessary under ordinary competitive situations. During the interval between inhalation and performance, the benefit of any increase in the oxygen saturation seems to be lost. (Sharkey, 1961.) However, the athlete is able to hyperventilate while on the starting block, and that procedure will lower the blood pCO_2, reducing his urge to breathe during the early portion of the race.

During performance Subjects were able to more than double their performance times by breathing 66 percent oxygen during exhaustive treadmill runs. Bannister and Cunningham (1954) reported that one subject quit only because he had to catch a train! Mountain climbers on the world's tallest peak, Mount Everest, have confirmed the value, indeed, the necessity of breathing oxygen during the arduous climb. However, no one has yet been able to devise a system that allows the athlete to carry extra oxygen *during* competition, and when someone does it will certainly be ruled a foul.

During recovery The use of oxygen as an aid to recovery presumes a more rapid resaturation of the blood hemoglobin and muscle myoglobin due to the elevated partial pressure of oxygen. In fact, even at the pO_2 of atmospheric air, the saturation returns to resting levels rather quickly. We observed no performance benefit when swimmers breathed oxygen prior to a second 100-yard race. (Sharkey, 1961.) The interval between races was 5 minutes. In another experiment involving near-maximal treadmill runs, we observed a small but interesting tendency toward faster heart-rate recovery between 5-minute runs. The interval in this experiment was but 1 minute. (Bjorgum and Sharkey, 1966.) It seems possible that recovery might be aided if the interval between maximal efforts is 1 minute or less. These findings should not be extrapolated to the brief anaerobic bursts of effort common to competition in football. The time spent in the huddle, in time-outs, and on the bench should be adequate for recovery under normal atmospheric conditions. It is likely that the athletes have been led to lean on the psychological benefit provided by the inhalation of the cool, rapidly expanding compressed gas.

At higher altitudes While it seems that the use of oxygen as a performance aid is entirely without justification at or near sea level, it may be of some benefit during recovery at higher altitudes (above 5000 feet). Since the pO_2 of the alveolar air declines with altitude, the recovery processes will take longer at higher elevations. Although research has not been conducted to determine the differences in recovery and the benefit of oxygen administration at higher altitudes, it is likely that the process would be aided by an increase in the alveolar pO_2. Such an increase would allow the repayment of the oxygen debt at a faster rate, and speed the saturation of hemoglobin and myoglobin.

CHAPTER 11 PHYSICAL ACTIVITY AND CARDIO-VASCULAR HEALTH

Along with mechanization, automation, the increased use of the automobile and other labor-saving devices has come a reduction in the requirement for physical activity and the greatest sustained epidemic that mankind has ever experienced—coronary heart disease (CHD). The greatest single cause of death in the United States and in many other highly developed countries, CHD victimizes over 600,000 Americans annually via the tragic episode known as the heart attack or *myocardial infarction.* The heart attack occurs when a blood clot or *thrombus* clogs a coronary artery already narrowed by *atherosclerosis.* Atherosclerosis results when fatty substances such as cholesterol are deposited beneath the lining of the arteries to form plaques. This deposition leads to a narrowing of the artery and a reduction of blood flow (ischemia). As the narrowing develops, the oxygen supply to the heart is reduced, its work capacity declines, and the risk of experiencing a heart attack grows.

For every case of known CHD (over 3 million), there is probably another yet to be diagnosed. Over 1 million heart attacks occur annually in this country. About half of the victims die in the first few hours. Among those experiencing their first episode, 25 percent die within 3 hours, another 10 percent within the week. Those who survive face the likehood that death will eventually come via the myocardial infarction (Braun and Diettert, 1972).

In this chapter we shall attempt to explore the relationship of physical activity to cardiovascular health in hopes of determining ways in which exercise can influence the factors involved in CHD—the atherosclerotic narrowing of the coronary arteries and the occlusion of these arteries with a clot.

RISK FACTORS

Numerous so-called risk factors have been statistically associated with the incidence of CHD. They include:

Influenced by physical activity
body type (endomesomorph)
overweight
elevated blood lipids (cholesterol and triglycerides)
physical inactivity

May be influenced by activity
carbohydrate intolerance
electrocardiographic abnormalities
elevated uric acid levels
pulmonary function (lung) abnormalities
personality or behavior pattern (hard-driving, time-conscious, aggressive, competitive)
psychic reactivity (reaction to stress)

Not influenced
family history of heart disease (heredity)
sex (male has greater risk until his sixties)
cigarette smoking
diet (coffee, sugar, saturated fats)

The coronary-prone individual is one who exhibits several of these risk factors; for example, the inactive smoker is a prime target for CHD. Participation in vigorous physical activity will reduce the risk to the level experienced by the inactive non-smoker. Remain active and give up smoking, and the risk is further reduced.

These risk factors have been *associated* with CHD. This means that there is a statistically significant relationship between the risk factor and the incidence of the disease. It *does not* imply cause and effect. It has been said that those who drink more than five cups of coffee daily are twice as likely to suffer heart attacks than those who drink no coffee at all. Is coffee the cause of CHD, or is it related to some other factor that may be the cause, such as psychic reactivity, behavior pattern, or even cigarette smoking? Similarly you may ask, does physical *in*activity cause heart disease, does it enhance the development of the disease, or is it merely related to some other causal factor?

PHYSICAL ACTIVITY AND CARDIOPROTECTION

Possible cardioprotective effects of exercise may be inferred from population studies of the incidence and mortality from CHD, or from postmortem investigations of atherosclerotic pathology. Data from numerous *population studies* typically indicate a lower incidence of CHD among the physically active members of the group in question, be they conductors and bus drivers, blue- and white-collar workers, farm workers and farm owners, or laborers and professional men. (Fox and Haskell, 1968.) The decline in the incidence of CHD ranges from 70–30 percent of the inactive rate as the level of activity increases (Fig. 11.1). Zukel, Lewis, and Enterline (1959) found that 1 hour of heavy

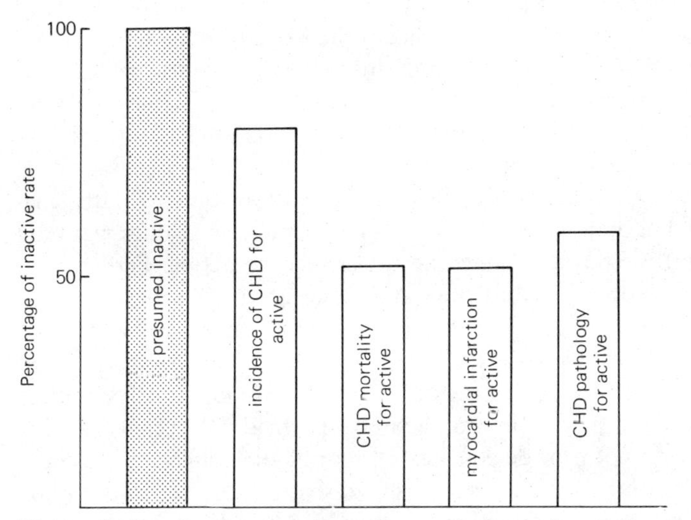

Figure 11.1. Cardioprotective effects of physical activity. Some of the cardioprotective tendencies noted in the literature are summarized here. The risk for active individuals is presented as a percentage of the risk for inactive subjects. The active individual has about one-half the risk of experiencing a myocardial infarction or heart attack as does his inactive neighbor.

physical activity daily lowered the incidence of CHD to less than 20 percent of the sedentary rate. The incidence of myocardial infarction is reduced by 50 percent or more, and CHD mortality may also be reduced to less than half of the sedentary rate, depending on the study chosen, the population investigated, and the level of physical activity involved. (Fox and Haskell, 1968.) These population studies do not prove that physical activity is the cause of the reduced incidence of CHD. Other factors may influence the selection of active or sedentary occupations, and factors associated with those occupations (such as stress, tension, responsibility) may influence the development of the disease. In fact, to illustrate how complicated the question is, Barry, Daly, and Kelly (1972) reported that the incidence of CHD is lower in those countries that spend more in support of the arts!

The *pathology* of CHD is significantly developed at the age of 22. (Enos, Beyer, and Holmes, 1955.) In a postmortem study of 300 American soldiers killed in the Korean conflict, 77 percent were found to have significant coronary artery disease. In another study, Morris and Crawford (1958) found that the incidence of fibrous patches, scars, infarcts, and occlusions was less frequent among those presumed to be physically active. However, the atherosclerosis or the deposition of fatty material (atheromas) in the wall of the coronary arteries was only slightly reduced among those classified as active. Perhaps our cardioprotective efforts should be concentrated on the school-age youngsters in an effort to minimize the early development of CHD pathology.

While the data suggest possible cardioprotective effects associated with physical activity, particularly *recent* activity, they do not provide conclusive proof of such effects. Such proof will await the results of yet to be conducted *intervention studies* that involve random assignment of human subjects to exercise and control groups for long-term observation of the effects of physical activity on the incidence of CHD. Studies of this kind are expensive to conduct, difficult to control, and subject to a considerable drop-out rate. Thus, it is possible that a truly conclusive study may be a long time in the making.

POSSIBLE CARDIOPROTECTIVE MECHANISMS

Although we lack conclusive evidence supporting the cardioprotective effects of exercise, there are several possible mechanisms whereby physical activity may reduce or delay the effects of CHD. (Fox, Naughton, and Garman, 1972.)

Physical activity may increase	*Physical activity may decrease*
coronary blood vessels	serum cholesterol and triglycerides
vessel size	glucose intolerance
efficiency of the heart	obesity, adiposity (fatness)
efficiency of peripheral blood distribution and return	platelet stickiness
fibrinolytic (clot-dissolving) capability	arterial blood pressure
arterial oxygen content	heart rate
	vulnerability to dysrhythmias

Physical activity may increase	*Physical activity may decrease*
red blood cells and blood volume	overreaction to hormones
thyroid function	strain associated with psychic stress
growth hormone production	
tolerance to stress	
prudent living habits	
the joy of living	

Let us consider some of these mechanisms and their relationship with various risk factors, atherosclerosis, and occlusion.

Myocardial Mechanisms

Physical activity improves the *efficiency* of the heart muscle and thereby lowers the oxygen needs of the myocardium. Raab (1965) suggested that exercise leads to an improvement in the autonomic regulation of the heart, a decrease in catecholamine-related oxygen consumption, and an improved myocardial electrolyte balance. The decrease in heart rate at rest and during submaximal exercise is vivid proof of improved myocardial efficiency.

We have already mentioned the increase in *coronary vascularization* that occurs with exercise. Stevenson et al. (1964) found that moderate physical activity was more effective in regard to coronary vascularization than was strenuous exercise.

Exercise may also enhance the intercoronary collateral circulation, the vessels that connect arterioles to allow alternative circulatory routes. (Eckstein, 1957.) Well-developed collateral circulation, in theory, would minimize the damage caused by a myocardial infarction, reduce the risk of death, and increase the chances for recovery from the episode. This theory is supported by data gathered on physically active populations. However, this increased collateral circulation has only been demonstrated in an already compromised circulation, where partial occlusion of the coronary vessels has already occurred.

Thus, collateralization may serve to relieve compromised (ischemic) heart tissue, minimize damage, and hasten recovery. Collaterals *do not* develop in the absence of exercise unless the circulation is severely impaired. The pathological symptoms of atherosclerosis are well developed in the early twenties so it would seem wise for all adults to engage in exercise as a form of cardioprotection.

Vascular Mechanisms

Regular, moderate physical activity seems to enhance the *fibrinolytic mechanism* which serves to dissolve intravascular clots. Fat seems to inhibit this mechanism and fibrinolytic activity seems to diminish as body fat increases. (Shaw and MacNaughton, 1963.) We have already mentioned that exhaustive, competitive, or unfamiliar exercise seems to accelerate clot formation.

Physical activity may also tend to reduce the incidence of *high blood pressure* (hypertension) among older individuals (Morris and Crawford, 1958) or middle-aged men (Boyer and Kasch, 1970).

Body Weight and Blood Lipids

While there may be no relationship between above-average *body weight* and heart disease when cases of hypertension and diabetes are excluded (Spain, Nathan, and Gellis, 1963), those of below average weight enjoy a lower rate of CHD. We will deal with the subject of exercise and weight control later in this section. Elevated *cholesterol* and *triglycerides* in the blood serum are risk factors that do not seem to behave consistently as a consequence of exercise. While one study showed a significant reduction in serum cholesterol following 1 week of intensive Alpine skiing (Doležel, 1968), another was unable to detect a change after 6 months of endurance exercises. (Skinner et al., 1967.) Serum triglycerides were reduced after the 6-month program, a reduction said to occur several hours after an exercise session and to last for about 2 days. Thus, if the reduction in serum triglycerides is an important cardioprotective mechanism, it would seem to require fairly regular physical activity to maintain the effect. Serum cholesterol only seems to be reduced in a heavy training program. (Rochelle, 1961.)

Psychological Factors

Selye (1956) suggested that an enjoyable interlude involving physical activity may improve our reaction to the stresses of life. If it is true that exercise improves our reaction to *psychic stress,* reduces tension, and alters the physiological manifestations associated with CHD, this aspect of exercise may be its most important contribution. Researchers at Purdue University have recently demonstrated the psychological benefits of physical activity on the personality profiles of unfit men (Ishmail and Trachtman, 1973).

It is clear that CHD is somehow related to life style, to a behavior pattern becoming ever more prevalent in our society. Friedman (1964) characterized a distinctive personality complex, behavior pattern A, involving extreme competitiveness, ambition, and a profound sense of time urgency. Men with type-A behavior had higher serum cholesterol levels and faster clotting times than their more relaxed counterparts. In spite of similar dietary and exercise habits, the type-A subjects had a sevenfold greater incidence of CHD.

Recently, Friedman reported an alarming number of sudden deaths among men diagnosed as type A. Half of the victims exercised strenuously within 6 hours of a large meal. He hypothesized that they may have carried the same competitiveness and time urgency into their leisure time. Some of the deaths were recorded after competitive handball and tennis matches, or after jogging. The deaths frequently occurred after a large meal, when serum lipid levels would inhibit fibrinolytic activity, and when the demands of digestion would seriously compromise the blood supply to the heart. In fairness to exercise it should be noted that half of the sudden deaths occurred while the victims were at rest or asleep. (Friedman and Rosenman, 1973.) Inactivity affords no protection from CHD.

Physical educators cannot ignore the potential dangers inherent in their programs, nor should they fail to understand the potential contributions of regular, moderate activity to cardiovascular health.

EXERCISE, PHYSIOLOGICAL FITNESS, AND WEIGHT CONTROL

For many years, the knowledge that carbohydrate was the major source of fuel for *vigorous* physical activity overshadowed the fact that fat was readily metabolized during light-to-moderate activity. Estimates of the exercise required to lose a pound of fat seemed excessive and impossible for all but a few hardy souls. For example, it has long been known that one must walk about 35 miles to lose 1 pound of fat (3500 calories). The simple fact that people failed to notice or mention is that one could also walk one mile a day for 35 days to accomplish the same goal. Let us now consider the ways in which exercise fitness and weight control are related and, as we develop these points, let us not forget the possible relationships between exercise, fitness, weight control—and cardiovascular health.

Exercise and Caloric Expenditure

Weight control (or energy balance) is achieved by balancing caloric intake and caloric expenditure. In a mechanized society this balance becomes difficult since occupational efforts have been reduced to minimal levels while automobile usage has increased. Furthermore, as we grow older we tend to maintain the dietary habits of our youth in spite of a gradual decline of metabolically active cells and basal energy requirements (about 4 percent for each decade over 25 years of age). Thus, caloric balance will eventually require either a reduction in the caloric intake and occasional hunger pangs, an increase in the caloric expenditure, or both.

By now you are well aware that exercise increases the caloric expenditure, and that the rate of expenditure is related to both the intensity and duration of the activity. Ten minutes of half-court basketball (utilizing about 7 calories per minute) will burn 70 calories; 20 minutes will consume 140, and 30 minutes will burn off about 210 calories above your basal energy expenditure. Thirty minutes of full-court basketball could burn off 360 calories or more, depending on the intensity of the game. Of course, as the exercise becomes more intense, we become limited in our ability to continue at that rate. That may explain the popularity of jogging, bicycling, and other moderate activities. While we may be able to expend as many as 125 calories in one all-out mile run (5 min $\times$ 25 calories per minute), we could jog at a comfortable pace for several miles (12 calories per minute) and triple the caloric expenditure (360 versus 125 calories) without becoming exhausted (Table 11.1).

Many would claim that diet, the reduction of caloric intake,

TABLE 11.1. THE CALORIC COST OF RUNNING[a]

Pace[b]	*Speed*	*Calories per Minute*
12-minute mile	5 MPH	10.0
8-minute mile	7.5 MPH	15.0
6-minute mile	10 MPH	20.0
5-minute mile	12 MPH	25.0

[a] Depends on efficiency and body weight. Use Fig. 12.1 to confirm your caloric expenditure. Adjust for weight (see Appendixes B and C).
[b] *Each* mile requires about 120 calories (Kcal) regardless of pace.

was the most effective means for weight control. Oscai and Holloszy (1969) studied the effects of diet and exercise on the body composition of laboratory rats. The experiment was controlled so that both groups lost the same amount of weight. The composition of the body substance lost via food restriction included 62 percent fat while the exercising animals lost 78 percent of their weight loss as fat.

	Exercise	*Caloric restriction*
fat	78%	62%
protein	5%	11%
minerals	1%	1%
water	16%	26%

The study provided vivid evidence of the protein-conserving and fat-metabolizing effects of physical activity.

Further proof of the value of exercise in weight loss via the oxidation of fat comes from an analysis of the effects of exercise on carbohydrate and fat metabolism. Paul and Issekutz (1967) studied the contributions of fat and carbohydrate as energy sources for the exercising dog. In short, intense effort carbohydrate supplied the major source of energy, but in prolonged activity, FFA provided the energy for contractions.

	Short, intense effort	*Prolonged effort*
FFA	20–30%	70–90%
glucose	70–80%	10–30%

Thus, it seems likely that fat is the major source of energy for light-to-moderate activity. Recent studies have suggested an important role for FFA metabolism over a wide range of work loads (Paul, 1970). It may be that fat metabolism declines only when the work load surpasses the ability of the circulation to supply FFA and oxygen, at which point anaerobic processes take over and glycogen stores serve as the major fuel.

While it has been established that exercise serves to metabolize fat as an energy source, we have not mentioned when that activity might be most beneficial. Zauner et al. (1971) have demonstrated the effect of exercise on the concentration of fat (lipid) substances in the blood. After fat is ingested it is absorbed via the lymphatic system and eventually appears in the blood for transport to active muscles or to adipose tissue. Moderate exercise appears to reduce both the magnitude and duration of this postprandial lipemia (i.e., the postmeal concentration of fat in the blood). The reduced incidence of CHD among habitually active populations may result in part from the influence of physical activity on blood lipid levels. This effect may be enhanced if the fat is metabolized before it has a chance to be deposited as an atherosclerotic plaque in the lining of an artery. The timing of the large meal and subsequent physical activity may be an important factor in the etiology of CHD.

Physiological Fitness and Weight Control

By now it is no secret that regular moderate physical activity may lead to an improvement in the aerobic capacity (see Table 11.2). This same sort of regular, moderate activity also seems to lead to the cardioprotective effects associated with physical activity. We shall use the term *physiological fitness* to differentiate the effects of exercise on the cardiorespiratory systems (associated with the aerobic capacity) from other interesting but unrelated aspects of physical or motor fitness, such as strength, speed, or flexibility. Physiological fitness enhances weight control and provides several exercise *extras*.

As we increase our aerobic capacity we are able to accomplish a given task with a lower heart rate. We perceive the task to be less demanding as cardiorespiratory efficiency improves. Thus,

TABLE 11.2. FITNESS COMPARISONS[a] (AEROBIC CAPACITY IN MILLILITERS PER KILOGRAM PER MINUTE)

Subjects	*Country*	*Men*	*Women*
untrained young	United States	43	30
untrained young	Canada	49	36
untrained young	Scandinavia	59	43
active young	United States	52	39
active young	Canada	55	—
active young	Scandinavia	59	—
champion distance athletes	United States	82	68
champion distance athletes	Sweden	82	68
untrained (40–50 yr)	United States	36	—
untrained (40–50 yr)	Canada	39	30
untrained (40–50 yr)	Scandinavia	45	34
trained (40–50 yr)	United States	58	—

[a] U.S. athletes seem as fit as those from other nations, but other members of the population suffer in comparison with those from countries where the life style includes a higher level of physical activity.
Sources: Shephard, 1966; Pollock, Miller, and Wilmore, 1972.

the fit individual is able to accomplish more work without a sense of fatigue. Moreover, since perception of effort is closely related to the exercise heart rate (Docktor and Sharkey, 1971) the fit individual may actually prefer to exercise at the same heart rate, thereby burning more calories per minute (see Fig. 12.1). In either case, physiological fitness allows a greater caloric expenditure without an increase in the sense of fatigue.

As one becomes fit he produces less lactic acid at a given work load, due to a greater reliance on aerobic energy sources. Issekutz and Miller (1962) reported that lactic acid seemed to inhibit the mobilization of FFA from fat storage deposits (see Fig. 8.2). Normally, epinephrine acts to stimulate fat mobilization from adipose tissue cells via activation of the fat-splitting enzyme lipase. When lactate levels rise, the release of FFA is inhibited. Physiological fitness leads to a reduction in lactic acid allowing an even greater mobilization of fat as a source of energy.

The mobilization of FFA does not ensure their metabolism. What of the effects of exercise, training, and fitness on the ability to oxidize FFA? Molé, Oscai, and Holloszy (1971) found that the ability to oxidize FFA was doubled in the muscles of rats following a period of endurance training. The authors suggested that the shift to fat metabolism during training was an important mechanism in the development of endurance fitness and an important factor serving to spare carbohydrate stores and prevent the hypoglycemia of prolonged exertion. Thus, the physiologically fit individual is able to derive a greater percentage of his energy requirements from fatty acid oxidation than is the unfit individual. At a given work load the fit individual may utilize as much as 90 percent fat. The *extra* benefits of physiological fitness for weight control and fat metabolism should be abundantly clear. By increasing energy requirements, one is able to burn more calories, mobilize greater amounts of fat, and improve fat utilization.

However, it is not yet clear how, and to what extent, physical activity may delay the onset or decrease the severity of CHD. We cannot say with assurance that a given amount of daily activity will provide adequate insurance against the manifestations of the nation's number one killer. We can say that regular, moderate physical activity influences several mechanisms that seem to be associated with a lower incidence of CHD.

CHAPTER 12 THE PRESCRIPTION OF EXERCISE

Physical activity may be an important factor in the attainment and maintenance of optimal health. As with any treatment or drug it must be used with respect if its benefits are to be realized and if we are to avoid its undesirable side effects. The amount of exercise that safely promotes the training effect leading to physiological fitness can be expressed in terms of specific training factors. Recent research has delineated some principles of exercise prescription that may someday lead to a pharmacopoeia of exercise.

THE TRAINING EFFECT

The possible role of exercise in the prevention or treatment of coronary heart disease (CHD) has shifted the attention of some exercise physiologists from maximal performance in athletics to optimal performance in the adult years. The research emphasis has shifted from a systems approach of physical training (e.g., distance running, interval training) to a search for specific factors associated with the training effect. In this section we shall consider those factors that seem to be involved in the training of physiological fitness or cardiorespiratory endurance (synonymous with aerobic capacity or maximal oxygen intake). Factors most frequently mentioned in the literature include the intensity, duration, and frequency of exercise as well as such individual factors as level of fitness, age, sex, and training goals.

Intensity of Exercise

We have made reference to exercise intensity throughout this monograph since it often determines the energy source utilized, the number of calories consumed, and the total duration of exercise. Intensity can be specified as a percentage of one's maximal oxygen intake (e.g., 50 percent of maximal intake), as the number of calories per minute (e.g., 10 calories per minute), or as a specific training heart rate (e.g., 150 bpm). While the first method of expression is appropriate for research purposes it is far too involved for practical usage. Thus, the exercise prescription should include reference to exercise intensity in terms of caloric consumption per minute or heart rate per minute.

Various studies have suggested the need for the exercise intensity to exceed a certain minimum if significant changes in cardiorespiratory endurance are to occur. Studies by Karvonen, Kentala, and Mustala (1957), and by Sharkey and Holleman (1967) and others agree that the heart rate must exceed 130 bpm (above 70 percent of the difference between the resting and maximal heart rate) in order to achieve a significant training effect. However, in these studies the increased intensity usually corresponded with an increase in the total work done. It appears possible that exercise of lesser intensity may elicit a training effect if continued for a sufficient duration. (Sharkey, 1970.)

The individual prescription of exercise intensity can incorporate a *training heart rate* (HR_T) based on 70 percent of the difference between the resting and maximal heart rates. (Karvonen, Kentala, Mustala, 1957.)

$$HR_T = \text{Rest HR} + 0.70\ (\text{Max HR} - \text{Rest HR})^*$$

For a man with a resting rate of 70 and a maximal rate of 170 bpm, the training heart rate will be 70 + 70 or 140 bpm. The training heart rate is roughly equivalent to a work level amounting to 70 percent of one's maximal oxygen intake. Athletes may want to train at higher intensities to hasten the achievement of maximal training benefits, but adults may safely exercise at this submaximal level to enjoy the benefits of physical activity and physiological fitness.

Duration of Exercise

Exercise duration can be prescribed in terms of time or distance, but when weight control is desired, the total oxygen cost or total caloric expenditure should be used as an indication of the length of the workout. Cardiorespiratory endurance has been improved in studies using 100-calorie workouts that last as little as 5 minutes (Bouchard et al., 1966; Sharkey, 1970). Cureton (1969) recommended 300–500 calorie workouts to gain fitness *as well as* the weight loss and fat metabolism benefits of exercise. The duration chosen will depend on the fitness of the individual, his training goals, and the exercise intensity employed.

Frequency of Exercise

While some studies have found two or three training sessions per week to be similar in effect to more frequent participation for subjects of low or average fitness (Jackson, Sharkey, and Johnston, 1968), Pollock, Cureton, and Greninger (1969) noted that improvements were in direct proportion to the frequency of training, especially in the latter stages of training. It may be that two or three training sessions per week are sufficient in the

* A simplified method utilizes 80 percent of the maximal heart rate or $0.80 \times 170 = 136$ bpm.

early weeks of training, but as the training progresses in intensity and duration, it must also increase in frequency. Those interested in weight control and fat metabolism benefits are encouraged to consider daily activity to maximize the caloric expenditure associated with physical activity.

Level of Fitness

Shephard (1968) and Sharkey (1970) have both noted the importance of the initial level of fitness as a consideration in individualized training prescriptions. Shephard reported that the response to a training program was largely determined by the intensity of effort relative to the individual's initial fitness level. We have found that changes in fitness were inversely related to the subject's initial level of fitness. In other words, individuals with higher initial fitness levels must work more to improve and can expect less overall improvement. Those of low fitness can look forward to considerable improvement. However, the exercise intensity should be scaled to fit their current exercise capacity. Those of low fitness should begin training at or near their training heart rates. Exercise duration need not exceed 100–200 calories at first, and two or three training sessions per week should be sufficient. Those of high fitness can tolerate high-intensity workouts in excess of 300 calories daily (Table 12.1).

Individual Factors

Since the *maximal heart rate* declines with age it is necessary to consider that fact in the calculation of the training heart rate and in the individualized prescription of exercise. Failure to make that adjustment could lead to serious medical consequences. (Wilmore and Haskell, 1971.) In his popular book, *Aerobics*, Cooper (1968) suggests a heart rate of 150 bpm as the level of intensity required to elicit a training effort. The 150-bpm heart rate would be close to the maximal rate for a man of 65 years. Since it would be imprudent to physically tax an older individual in order to determine his maximal heart rate, we have included average maximal heart rates for use in the determination of the training heart rate. [Recent evidence from Pollock, Miller, and Wilmore (1972) suggests that the

TABLE 12.1. PHYSIOLOGICAL FITNESS: SUGGESTED EXERCISE PRESCRIPTIONS TO INITIATE THE FITNESS PROGRAM[c]

Fitness (milliliters per kilogram per minute)	*Intensity*[a]	*Duration (calories)*	*Frequency*
Low <35	<HR_T[b]	100	Alternate days
Medium 35–45	HR_T	200–300	Alternate Days→daily
High >45	>HR_T	>300	Daily→ Twice daily

[a] Or 80 percent of maximal heart rate.

[b] HR_T denotes training heart rate.

[c] The average woman weighs less, has a bit more body fat, has somewhat less hemoglobin and about three-fourths the aerobic capacity of the average man. Thus some minor adjustments are necessary to adapt the exercise prescription for feminine use.

Intensity: use HR_T; no adjustment needed

Duration: women are smaller, hence they burn fewer calories in a given activity. Adjust total caloric expenditure by subtracting 25 percent.

$100 - 25 = 75$ calories per workout
$200 - 50 = 150$ calories per workout
$300 - 75 = 225$ calories per workout

Frequency: no adjustment needed.

maximal heart rates of chronically active senior athletes are less affected by age; see page 182.]

One's exercise interests and training goals may be related to his or her *sex.* Men may be more interested in achieving the cardioprotective effects of exercise while women, seemingly protected from CHD until the age of menopause, may be more interested in appearance and weight control benefits. Since it is impossible to say how much exercise is necessary for optimal health, individual exercise goals and optimal levels of cardiorespiratory endurance (physiological fitness) require very personal definition. A bowler or golfer will be satisfied with a lower level of fitness than a tennis player or a backpacker. One man

Age (years)	*Maximal Heart Rate (bpm)*	HR_T[a]
20–30	190	152
30–40	180	144
40–50	170 (178)[b]	136
50–60	160 (175)	128
60–70	150 (163)	120
70–80	140	112

[a] 80 percent of maximal HR_T.
[b] American champion track athletes.

may continue to set ever-higher goals in jogging, while another may decide to quit when he reaches a desired level of fitness. One young woman may bicycle to prepare for the ski season while another may ride solely for the pleasure it provides.

Mode of Exercise

Recently, Pollock and associates (1972) compared the fitness and weight control benefits of three popular modes of exercise: walking, running, and cycling. Sedentary middle-aged men trained at the same intensity, duration, and frequency of exercise for 20 weeks. All three groups improved similarly in cardiorespiratory fitness, body weight, and skinfold fat. When intensity, duration, and frequency of exercise are similar, the benefits of exercise are similar.

PHYSIOLOGICAL FITNESS

The exercise prescription designed to improve physiological fitness may involve a wide range of intensities and durations of exercise, depending on the fitness, interests, and goals of the individual. If your goal is to improve fitness for general health and vitality, your intensity and duration of exercise may be rather modest. However, if you are training for an extended hike or ski tour you will want to place a sufficient overload on your cardiorespiratory system in order to maximize the training effect. When training for fitness you may be satisfied with two or three sessions per week, unless you have reached the point when further improvement requires more frequent training periods (see Table 12.1). Whatever your intensity, duration, or frequency of training, you should not attempt to rush the training program. Avoid the urge to do just a little more, to push just a little harder.

Perhaps the best advice might be to *integrate* your fitness efforts with activities you enjoy. If you find that you are unable to get sufficient activity in that manner, *supplement* your game of volleyball with some jogging and warming-up exercises. Supplements may also be desirable to aid your improvement in a given activity (e.g., leg strength for skiing). In any case, you should avoid the compulsive drive to remain on an unrewarding exercise regime. Fitness for fitness' sake is an empty goal. Pursue fitness for its benefits if you must, but consider also how it allows you to adapt, how it adds life to your years, not just years to your life.

Once achieved, a desired level of fitness may be *maintained* with one or two exercise sessions per week. The intensity of the exercise must be comparable to that used to achieve the current level of fitness. A considerable amount of your hard-earned improvement may be lost if you decide to refrain from exercise for an extended period of time. On the other hand, you are not likely to refrain from exercise once it has become a part of your life style. The phenomenal growth in the use of the bicycle as a means of transportation illustrates how some have integrated physical activity into their life style; at the same time they help to curb air pollution while seeking a quiet interlude at the beginning and end of the workday.

WEIGHT CONTROL

The exercise prescription designed to achieve weight loss or weight control must maximize caloric expenditure at the expense of exercise intensity. Duration and frequency of activity must be increased to achieve higher levels of caloric expenditure and subsequent weight loss. As the exercise prescription is best based on a knowledge of one's fitness (Table 12.2), the weight control prescription should be based on facts concerning the caloric balance.

Caloric Intake

You may figure your caloric intake by keeping records of all the food you eat, including snacks, and then figuring the calories per serving, per meal, and per day from a calorie chart.

TABLE 12.2. WEIGHT CONTROL: SUGGESTED EXERCISE PRESCRIPTIONS TO INITIATE THE WEIGHT CONTROL PROGRAM

Fitness (milliliters per kilogram of body weight per minute)	*Intensity*[a]	*Duration (calories)*	*Frequency*
Low <35	$<HR_T$[b]	100–200	Alternate day—daily
Medium 35–45	HR_T	200–400	Daily
High >45	$>HR_T$	>400	Daily—Twice daily or long duration

[a] Or 80 percent of maximal heart rate.
[b] HR_T denotes training heart rate.

This is a most educational process and we should all do it now and then. With luck you might find someone to predict your percentage of body fat with skinfold calipers. You could then decide if you needed to remove some unwanted fat. Perhaps the easiest way to figure your caloric balance is to step on a scale and compare your weight with that listed in the desirable weight chart. (Table 12.3). If you are more than 10 percent above your "desirable" body weight you are overweight and more likely to suffer from CHD, diabetes, or hypertension. A quick glance in the mirror will convince some of you of the need to do something about your weight. Being *overweight* means there is an excess of fat *or* muscle, while being *obese* means there is an excessive accumulation of fat (20 percent fat for young men; 30 percent fat for women).

Caloric Expenditure

Since weight is gained when the caloric intake exceeds the expenditure, the methods listed above should shed some light on your average caloric expenditure. Caloric expenditure can

TABLE 12.3. DESIRABLE BODY WEIGHTS[a]

Height in Inches	*Weight in Pounds*	
	Men	*Women*
60	—	109 + 9[b]
62	—	115 + 9
64	133 + 11	122 + 10
66	142 + 12	129 + 10
68	151 + 14	136 + 10
70	159 + 14	144 + 11
72	167 + 15	152 + 12
74	175 + 15	—
76	182 + 16	—

[a] Heights and weights are without shoes and other clothing.
[b] Desirable weight for a small-framed woman of this height would be approximately 109 lb minus 9 lb, or a total of 100 lb; for an average-framed woman, 109 lb; for a large-framed woman, 118 lb.
Source: Food and Nutrition Board, National Research Council.

be estimated via a careful recording of *all* the day's activities, including sleep, followed by reference to elaborate tables listing the caloric cost (per minute) of a wide variety of household, industrial, and physical activities (see Appendixes B and C). A short method for the assessment of the energy expenditure is also included in Appendix A. Your weight control activities should be selected in terms of their contribution to a high caloric expenditure (Table 12.4). They should be of moderate intensity, and should be continued long enough to achieve a total expenditure of 300 calories or more. Daily exercise sessions would be most desirable. If caloric intake remains the same and the exercise expenditure amounts to 300 calories a day for 6 days, you will lose a pound of fat in about 2 weeks ($2 \times 1800 = 3600$ calories). As fitness improves, and you are able to burn even more calories, the extra benefits of exercise become more evident.

THE PROGRAM

If you have been inactive for some time you may want to see your physician for a medical examination before you undertake

TABLE 12.4. PHYSICAL ACTIVITY AND CALORIC EXPENDITURE

Work Intensity	*Pulse Rate*	*Calories (per minute)*	*Examples*
Light	Below 120	Under 5	Golf, bowling, Walking, volleyball, most forms of work
Moderate[a]	120–150	5–10	Jogging, tennis, bike riding, handball, basketball, hiking, strenuous work
Heavy	Above 150	Above 10	Running, fast swimming, other brief and intense efforts

[a] Preferred for weight-control benefits.

a fitness or weight control program. However, if you plan to remain *inactive*, the medical examination may be even more important. You may wish to calculate your caloric intake and expenditure, and your caloric balance before you begin. You may estimate your physiological fitness or aerobic capacity with the 15-minute run prediction of the maximal oxygen intake (Fig. 9.3). This test is *not* recommended for previously inactive or older individuals. Submaximal prediction methods are available for less active individuals (Sharkey, 1974). Prior to the start of your program it is wise to record body weight, resting pulse rate, and other information for future comparisons.

The program record should list types of activities or exercises, exercise intensity (training heart rate), duration, frequency, and desired caloric expenditure. Begin your participation gradually to avoid soreness or injury. In the event of soreness, deVries (1974) recommends static stretching as an effective method to achieve relief. You may check your caloric expenditure in any given activity by referring to Appendixes B and C. A personalized check of your expenditure may be made by recording your heart rate during the activity since the energy expenditure is directly related to the heart rate. Simply pause long enough to take a 10- or 15-second pulse count, determine the rate per

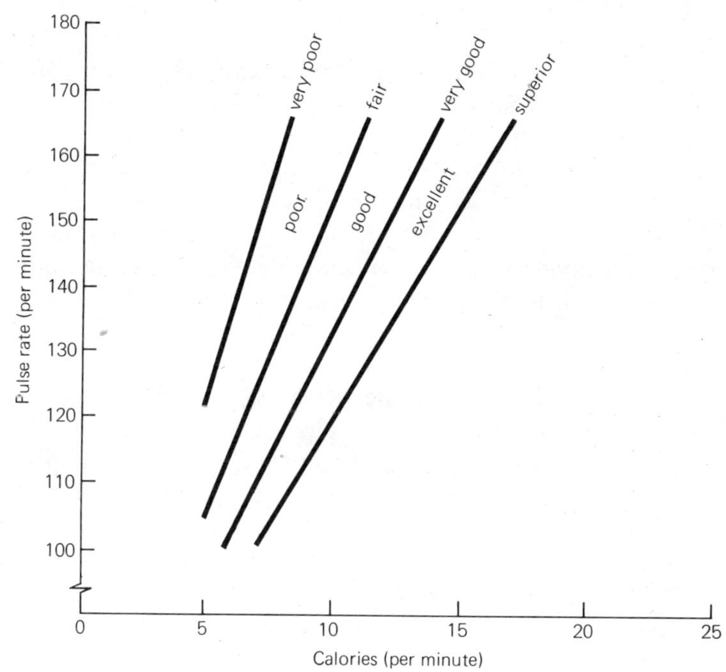

Figure 12.1. Predicting the calories burned during physical activity. Use a fifteen-second pulse count taken *immediately* after exercise (15-second rate × 4 = rate per minute). Then use the line corresponding to your level of fitness (see Fig. 9.3) and find your caloric expenditure. A rate of 150 would predict almost 10 calories per minute for one in the fair category, and almost 15 calories per minute for one in the superior fitness category. (For a 150-pound individual: add 10 percent for *each* 15 pounds over 150; subtract 10 percent for each 15 pounds under 150 (e.g., 165 lbs, pulse = 140, good category = 10 calories + 1, or 10 + 10 percent of 10 = 11 calories per minute).)

minute, and consult Fig. 12.1 at the line indicating your approximate fitness level.

Your daily program should include:

1. *Warm-up:* a good time for flexibility exercises
2. *Conditioning exercises:* optional muscular fitness activities (strength, endurance)
3. *Cardiorespiratory activities:* the core of the fitness program

In addition to a periodic check of your resting and exercise heart rates, your body weight, activity program, and exercise goals, it is often helpful to develop a *seasonal activity plan*. Write down your favored activities for each season. Add other types of activities, sports, and exercise supplements. Then analyze the plan to see where gaps appear in your year-round program. You may find a month or two in the fall when you tend to spend too much time viewing and too little time doing. You could decide to develop a preseason ski-conditioning program that will help you to improve your skill on the slopes. Or you could make the season more enjoyable by taking up hunting or bicycling in the fall, and still improve your fitness for winter sports. The most effective fitness and weight control programs are those that integrate physical activity into the life style. Run, jog, do calisthenics if you wish, but do not stop at that. The world of physical activity has so much to offer it would be a shame to limit your participation to its elementary forms. Try hiking, ski touring, bicycling; take up golf, tennis, or swimming again. Plan your program and get started, you will not regret it.

PRECAUTIONS

A final note is in order before we conclude our treatise on the physiology of physical activity. Throughout this monograph we have attempted to show both sides of exercise, the good and the bad. We have noted the various effects of training as well as some potential cardioprotective mechanisms. We have also alluded to the dangers inactive adults may face in highly competitive, exhaustive, threatening, or unfamiliar exertion. It is important that you carefully consider the problems involved in the prescription of exercise to adults.

In 1971, sports fans were shocked to see an apparently healthy young professional football player collapse and die during a nationally televised football game. Jokl and McClellan (1970) have reported similar cases of cardiac arrest in equally fit, active, and symptom-free individuals ranging from 7–45 years of age. Bruce and Kluge (1971) have been conducting closely supervised training programs for patients with clinically established coronary heart disease. They have reported seven cases of exercise-related cardiac fibrillation. Two cases occurred during exercise testing and five during the training program. All of the patients recovered following cardiac defibrillation, and six of the seven resumed physical activity within a few weeks of the episode. Bruce and Kluge suggest that physicians inform patients of the risk of exertional cardiac arrest and provide appropriate emergency facilities for the testing and training of coronary patients. They also recommend that participation in vigorous training programs be restricted to professionally supervised group programs. Cooper (1970) has published a list of precautions to observe in medically supervised exercise programs. They include a wait of at least 2 hours between mealtime and exercise, the avoidance of maximal effort during extreme weather conditions, an active recovery following exertion, and sufficient recovery and cool-down time before showering. He also recommends that a daily log be kept to record details concerning the day's exercise as well as the occurrence of symptoms or warning signs.

It is clear that the prescription of exercise to coronary patients is a medical problem requiring strict supervision as well as appropriate emergency equipment. But that does not suggest that these individuals need to avoid activity. In fact, 7 former heart attack victims completed the 1973 Boston marathon (26 miles, 385 yds). Yes, even the post-coronary patient benefits from a carefully administered exercise prescription.

Otherwise healthy, but sedentary, individuals should not be accepted into active exercise programs until they have received a complete medical examination. The American Heart Association (1972) recommends an examination as well as a resting electrocardiogram for those 30–40 years of age, and an exercise ECG for those over 40 years. The physical educator should require adherence to these guidelines to protect the health of program participants.

We should be prudent in the prescription of exercise, and our prescription should be viewed only as a guide. The exercising adult is always free to exert himself more or less—as he desires. We do not wish to force him to work until exhausted, nor do we want to limit the scope and imaginative nature of his involvement. The goal of the exercise prescription should be to provide him with the fitness and energy he needs to engage fully in life. The best physical educator is the one who provides experiences, information, encouragement and guidance, and eventually sends his student forth with the knowledge needed to seek a creative adaptation to life.

APPENDIXES

APPENDIX A. Assessment of Energy Expenditure

(Follow steps 1 through 4)

1. Calculate basal energy expenditure:

BASAL ENERGY EXPENDITURE FOR MEN AND WOMEN[a]

Men		*Women*	
Weight	*Caloric Expenditure*[b]	*Weight*	*Caloric Expenditure*[c]
140	1550	100	1225
160	1640	120	1320
180	1730	140	1400
200	1815	160	1485
220	1900	180	1575

[a] Basal energy = calories expended in 24 hours of complete bed rest.
[b] 5′10″ tall (add 20 calories for each inch taller, if shorter subtract 20 calories).
[c] 5′6″ tall (add 20 calories for each inch taller, if shorter subtract 20 calories).

2. Add increases in caloric expenditure:

APPROXIMATE INCREASES IN CALORIC EXPENDITURE FOR SELECTED ACTIVITIES

Activity	*Percent Above Basal*
bed rest (eat and read)	10 percent
quiet sitting (read, knit)	30 percent
light activity (office work)	40–60 percent
moderate activity (housewife)	60–80 percent
heavy occupational activity (construction)	100 percent

3. Adjust total for age:
   Subtract 4 percent of caloric expenditure for each decade (ten years) over twenty-five years of age.

4. Add calories expended in nonwork (recreational) activities:
   Use caloric expenditure charts in Appendixes B and C. Figure minutes of activity and cost in calories per minute. For example, a 5′10″, 200-pound, forty-five-year-old construction worker:

Basal = 1815 + 100 percent = 3630 − 8 percent (age) = 3340
Table tennis (30 minutes × 5) = 150

Total = 3490

APPENDIX B. Caloric Expenditure During Various Activities

Activity	Calories per Minute[a]
sleeping	1.2
resting in bed	1.3
sitting, normally	1.3
sitting, reading	1.3
lying, quietly	1.3
sitting, eating	1.5
sitting, playing cards	1.5
standing, normally	1.5
classwork, lecture (listen to)	1.7
conversing	1.8
personal toilet	2.0
sitting, writing	2.6
standing, light activity	2.6
washing and dressing	2.6
washing and shaving	2.6
driving a car	2.8
washing clothes	3.1
walking indoors	3.1
shining shoes	3.2
making bed	3.4
dressing	3.4
showering	3.4
driving motorcycle	3.4

Activity	*Calories per Minute*[a]
metal working	3.5
house painting	3.5
cleaning windows	3.7
carpentry	3.8
farming chores	3.8
sweeping floors	3.9
plastering walls	4.1
truck and automobile repair	4.2
ironing clothes	4.2
farming, planting, hoeing, raking	4.7
mixing cement	4.7
mopping floors	4.9
repaving roads	5.0
gardening, weeding	5.6
stacking lumber	5.8
stone, masonry	6.3
pick-and-shovel work	6.7
farming, haying, plowing with horse	6.7
shoveling (miners)	6.8
walking downstairs	7.1
chopping wood	7.5
gardening, digging	8.6
walking upstairs	10.0

[a] Depends on efficiency and body size. Add 10 percent for each 15 lb above 150, subtract 10 percent for each 15 lb under 150. Use activity pulse rate to confirm the caloric expenditures (Figure 12.1).

APPENDIX C. Calories Expended in Various Physical Activities

Activity	Calories per Minute[a]
pool or billiards	1.8
canoeing: 2.5 mph—4.0 mph	3.0–7.0
volleyball: recreational—competitive	3.5–8.0
golf: foursome—twosome	3.7–5.0
horseshoes	3.8
baseball (except pitcher)	4.7
ping pong—table tennis	4.9–7.0
calisthenics	5.0
rowing: pleasure—vigorous	5–15
cycling: 5–15 mph (10 speed)	5–12
skating: recreation—vigorous	5–15
archery	5.2
badminton: recreational—competitive	5.2–10
basketball: half—full court (more for fast break)	6–9
bowling (while active)	7.0
tennis: recreational—competitive	7–11
water skiing	8.0
soccer	9.0
snowshoeing (2.5 mph)	9.0
handball and squash	10.0
mountain climbing	10.0
judo and karate	13.0
football (while active)	13.3
wrestling	14.4

Activity	Calories per Minute[a]
skiing: moderate to steep	8–12
downhill racing	16.5
cross-country: 3–8 mph	9–17
swimming: pleasure	6.0
crawl: 25–50 yds per min	6–12.5
butterfly: 50 yds per min	14.0
backstroke: 25–50 yds per min	6–12.5
breaststroke: 25–50 yds per min	6–12.5
sidestroke: 40 yds per min	11.0
dancing: modern: moderate—vigorous	4.2–5.7
ballroom: waltz—rumba	5.7–7.0
square	7.7
walking: road—field (3.5 mph)	5.6–7.0
snow: hard—soft (3.5–2.5 mph)	10–20
uphill: 5—10—15 percent (3.5 mph)	8–11–15
downhill: 5—10 percent (2.5 mph)	3.6–3.5
15—20 percent (2.5 mph)	3.7–4.3
hiking: 40 lb pack (3.0 mph)	6.8
running: 12-minute mile (5 mph)	10.0
8-minute mile (7.5 mph)	15.0
6-minute mile (10 mph)	20.0
5-minute mile (12 mph)	25.0

[a] Depends on efficiency and body size. Add 10 percent for each 15 lb over 150, subtract 10 percent for each 15 lb under 150. Use activity pulse rate to confirm the caloric expenditure (see Fig. 12.1).

Sources: Passmore and Durnin, 1955; Consolazio, Johnson, and Pecora, 1963; Roth, 1968; Human Performance Lab, University of Montana, 1964–1974.

REFERENCES

Armstrong, R. B.; Saltin, B.; and Gollnick, P. D. 1973. Changes in human skeletal muscle fibers during exercise and after training. Paper read at the annual meeting of the American College of Sports Medicine, Seattle, Washington.

Astrand, P. O., and Rodahl, K. 1970. *Textbook of Work Physiology*. New York: McGraw-Hill.

Baldwin, K. M.; Klinkerfuss, G. H.; Terjung, R. L.; Molé, P. A.; and Holloszy, J. O. 1972. Respiratory capacity of white, red, and intermediate muscle: Adaptive response to exercise. *American Journal of Physiology* 222, 373–378.

Balke, B. 1963. A simple field test for the assessment of physical fitness. Report 63–6. Oklahoma City: Civic Aeronautic Research Institute, Federal Aviation Agency.

———; Daniels, J. T.; and Faulkner, J. A. 1967. Training for maximum performance at altitude. In Margaria, R. *Exercise at Altitude*. Amsterdam: Excerpta Medica, pp. 179–188.

———. 1968. Variation in altitude and its effects on exercise performance. In Falls, H. B. (ed.). *Exercise Physiology*. New York: Academic, pp. 240–267.

Bannister, R. G., and Cunningham, D. J. 1954. The effects, on the respiration and performance during exercise, of adding oxygen to the inspired air. *Journal of Physiology* 125, 118–135.

Barnard, R. J.; Edgerton, V. R.; and Peter, J. B. 1970. Effect of exercise on skeletal muscle I. Biochemical and histochemical properties. *Journal of Applied Physiology* 28, 762–765.

Barry, A.; Daly, J.; and Kelly, J. 1972. Cognitive function and coronary heart disease. Paper presented at the annual meeting of the American College of Sports Medicine, Philadelphia.

Bates, D. V. 1972. Short-term effects of ozone on the lung. *Journal of Applied Physiology* 32, 176–181.

Berger, R. A. 1962. Optimum repetitions for the development of strength. *Research Quarterly* 33, 334–338.

Bergström, J. 1962. Muscle electrolytes in man. *Scandinavian Journal of Clinical Laboratory Investigation* (14 Suppl.) 68.

Bjorgum, R. K., and Sharkey, B. J. 1966. Inhalation of oxygen as an aid to recovery after exertion. *Research Quarterly* 37, 462–467.

Bouchard, C.; Hollmann, W.; Venrath, H.; Herkenrath, G.; and Schlussel, H. 1966. Minimal amount of physical training for the prevention of cardiovascular diseases. Paper read at the 16th World Conference for Sports Medicine, Hanover, Germany.

Bourne, G. H. Nutrition and exercise. 1968. In Falls, H. B. (ed.). *Exercise Physiology*. New York: Academic, pp. 155–172.

Bowerman, W. J., and Harris, W. E. 1967. *Jogging*. New York: Grosset & Dunlap.

Boyer, J. L., and Kasch, F. W. 1970. Exercise therapy in hypertensive men. *Journal of the American Medical Association* 211, 1668–1671.

Braun, H. A., and Diettert, G. A. 1972. Coronary Care Unit Nursing. Missoula, Mont.: Mountain Press.

Brose, D. E., and Hanson, Dale L. 1967. Effects of overload training on velocity and accuracy of throwing. *Research Quarterly* 38, 528–533.

Brown, C. H., and Wilmore, J. H. 1971. Physical and physiological profiles of champion women long-distance runners. Paper presented at the annual meeting of the American College of Sports Medicine, Toronto.

Bruce, R. A., and Kluge, W. 1971. Defibrillatory treatment of exertional cardiac arrest in coronary disease. *Journal of the American Medical Association* 216, 653–658.

Burke, R. F.; Levine, D. N.; and Zajac, F. E. 1971. Mammalian motor units: Physiological-histochemical correlation in three types of cat gastrocnemius. *Science* 174, 709–712.

Buskirk, E. R. 1971. Work and fatigue in high altitude. In Simonson, E. (ed.). *Physiology of Work Capacity and Fatigue*. Springfield, Ill.: Thomas, pp. 312–323.

———, and Haymes, E. M. 1972. Nutritional requirements for women in sport. In Harris, D. V. (ed.). *Women in Sport*. Proceedings of a national research conference published by the

College of Health, Physical Education, and Recreation, The Pennsylvania State University, University Park, Pennsylvania.

Byers, S. O.; Friedman, M.; Rosenman, R. H.; and Freed, S. C. 1962. Excretion of 3-methoxy-4-hydroxymandelic acid in men with behavior pattern associated with high incidence of coronary artery disease. *Federation Proceedings* 21, 99–101.

Casner, S. W.; Early, R. G.; and Carlson, B. R. 1971. Anabolic steroid effects on body composition in normal young men. *Journal of Sports Medicine* 11, 98–103.

Cavagna, G. A.; Dusman, B.; and Margaria, R. 1968. Positive work done by a previously stretched muscle. *Journal of Applied Physiology* 24, 21–32.

Cherniack, R. M., and Cherniack, L. 1962. *Respiration in Health and Disease*. Philadelphia: Saunders.

Christensen, E. H., and Hansen, O. 1939. Arbeitsfähigkeit und shrnährung *Working Capacity and Diet. Scandinavian Archives of Physiology* 81, 160–172. (in German)

Clarke, H. H. (ed.). 1972. *Physical fitness research digest*. Washington, D.C.: President's Council on Physical Fitness and Sports, 2 (April, July and October).

Comroe, J. H. 1965. *Physiology of Respiration*. Chicago: Year Book Medical.

Connell, A. M.; Cooper, J.; and Redfearn, J. W. 1958. The contrasting effects of emotional tension and physical exercise on the excretion of 17-ketogenic steroids and 17-ketosteroids. *Acta Endocrinology* 27, 179–194.

Consolazio, C. F.; Johnson, R. E.; and Pecora, L. J. 1963. *Physiological Measurements of Metabolic Functions in Man*. New York: McGraw-Hill.

Cooper, K. 1968. *Aerobics*. New York: Bantam.

———. 1970. *The New Aerobics*. New York: Bantam.

———. 1970. Guidelines in the management of the exercising patient. *Journal of the American Medical Association* 211, 1663–1667.

Costill, D. L.; Miller, S. J.; Myers, W. C.; Kehoe, F. M.; and Hoffman, W. M. 1968. Relationship among selected tests of explosive leg strength and power. *Research Quarterly* 39, 785–787.

———**; Kammer, W. F.; and Fisher, A.** 1970. Fluid ingestion during distance running. *Archives of Environmental Health* 21, 520–525.

———**; Saltin, B.; Söderberg, M.; and Jansson, L.** 1973. Factors limiting the ability to replace fluids during prolonged exercise. Paper read at the annual meeting of the American College of Sports Medicine, Seattle, Washington.

Cumming, G. R. 1971. Correlation of physical performance with laboratory measures of fitness. In Shephard, R. J. (ed.). *Frontiers of Fitness*. Springfield, Ill.: Thomas, pp. 265–279.

Cunningham, D. A., and Faulkner, J. A. 1969. The effect of training on aerobic and anaerobic metabolism during a short exhaustive run. *Medicine and Science in Sports* 1, 65–69.

Cureton, T. K. 1969. *The Physiological Effects of Exercise Programs Upon Adults*. Springfield, Ill.: Thomas.

Dainty, D. 1971. The relationship of anaerobic capacity to selected performance tests. Master's thesis. University of Montana, Billings.

Daniels, J., and Oldridge, N. 1971. Changes in oxygen consumption of young boys during growth and running training. *Medicine and Science in Sports* 3, 161–165.

———. 1972. Personal communication.

Deutsch, J. A. 1971. The cholinergic synapse and the site of memory. *Science* 174, 788–794.

deVries, H. A. 1974. *Physiology of Exercise*. Dubuque, Iowa: Brown.

Dill, D. B., and Adams, W. C. 1971. Maximal oxygen uptake at sea level and at 3090 meter altitude in high school champion runners. *Journal of Applied Physiology* 30, 854–859.

Docktor, R., and Sharkey, B. J. 1971. Note on some physiological and subjective reactions to exercise and training. *Perceptual and Motor Skills* 32, 233–234.

Doležel, J. 1968. The effect of two types of physical strain during summer and winter on cholesterolemia in young people. In Poortmans, J. R. (ed.). *Biochemistry of Exercise*. Baltimore: University Park, pp. 148–151.

Donevan, R. E.; Palmer, W. H.; Varvis, C. J.; and Bates, D. V. 1959. Influence of age on pulmonary diffusing capacity. *Journal of Applied Physiology* 14, 483–492.

Eccles, J. C. 1964. *The Physiology of Synapses*. New York: Academic.

Eckstein, R. 1957. Effect of exercise and coronary artery narrowing on coronary collateral circulation. *Circulation Research* 5, 230–238.

Ekblom, B.; Goldbarg, A.; and Gullbring, B. 1973. Response to exercise after blood loss and reinfusion. *Journal of Applied Physiology* 35, 175–180.

———**, and Hermansen, L.** 1968. Cardiac output in athletes. *Journal of Applied Physiology* 25, 619–625.

Enos, W. F.; Beyer, J. C.; and Holmes, R. H. 1955. Pathogenesis of coronary disease in American soldiers killed in Korea. *Journal of the American Medical Association* 158, 912–914.

Folk, G. E. 1974. *Environmental Physiology*. Philadelphia: Lea & Febiger.

Fowler, W. H.; Gardner, G. H.; and Egstrom, G. H. 1965. Effect of an anabolic steroid on physical performance of young men. *Journal of Applied Physiology* 20, 1038–1040.

Fox, S. M., and Haskell, W. L. 1968. Physical activity and the prevention of coronary heart disease. *Bulletin of the New York Academy of Sciences* 44, 950–967.

———; Naughton, J. P.; and Garman, P. A. 1972. Physical activity and cardiovascular health. *Modern Concepts of Cardiovascular Disease* 1, 17–20.

Frankenhauser, M.; Post, B.; Nordheden, B.; and Sjoeberg, H. 1969. Physiological and subjective reactions to different physical work loads. *Perceptual and Motor Skills* 28, 343–349.

Frenkl, R., and Caslay, L. 1962. Effect of regular muscular activity on adrenocortical function in rats. *Journal of Sports Medicine* 2, 207–211.

Friedman, M. 1964. Behavior pattern and its pathogenetic role in clinical coronary artery disease. *Geriatrics* 19, 562–567.

———, and Rosenman, R. 1973. Instantaneous and sudden death. *Journal of the American Medical Association* 22, 1319–1328.

Gale, J. B. 1970. Skeletal muscle changes in ATP, CP, DPNH-reductase, and phosphorylase in rats trained at 900 and 7600 feet altitude. Doctoral dissertation, University of Wisconsin, Madison.

Ganong, W. F. 1963. Secretion and release of ACTH. In Nalbandov, A. (ed.). *Advances in Neuroendocrinology*. Urbana: Univ. of Illinois Press, pp. 107–120.

———. 1971. *Review of Medical Physiology*. Los Altos, Calif.: Lange.

Gollnick, P. D. 1971. Cellular adaptations to exercise. In Shephard, R. J. (ed.). *Frontiers of Fitness*. Springfield, Ill.: Thomas, pp. 112–128.

———; Armstrong, R. B.; Saltin, B.; and others. 1973. Effect of training on enzyme activity and fiber composition of human skeletal muscle. *Journal of Applied Physiology* 34, 107–111.

———, and King, D. W. 1969. Effect of exercise and training on mitochondria of rat skeletal muscle. *American Journal of Physiology* 216, 1502–1509.

———; Piehl, K.; Saubert, C. W.; Armstrong, R. B.; and Saltin, B. 1972. Diet, exercise and glycogen changes in human muscle fibers. *Journal of Applied Physiology* 33, 421–425.

Gordon, E. E. 1967. Anatomical and biochemical adaptations of muscle to different exercises. *Journal of the American Medical Association* 201, 755–758.

Greenleaf, J. E., and Castle, B. L. 1971. Exercise temperature regulation in man during hypohydration and hyperhydration. *Journal of Applied Physiology* 30, 847–853.

Guth, L. 1968. Trophic influences of nerve on muscle. *Physiological Review* 48, 645–680.

Gutmann, E. 1962. *The Denervated Muscle*. Prague: Czechoslovak Academy of Science.

Guyton, A. C. 1964. *Function of the Human Body*. Philadelphia: Saunders.

Haberern, J. (ed.) 1974. *Fitness for Living*. Emmaus, Pa.: Rodale Press (a bimonthly publication available by subscription or on the newsstand).

Hartley, L. H.; Mason, J. W.; Hogan, R. P.; and others. 1972. Multiple hormonal responses to graded exercise in relation to physical training. *Journal of Applied Physiology* 33, 602–610.

Haymes, E. M.; Harris, D. V.; Beldon, M. D.; and others. 1972. The effect of the physical activity level on selected hematological variables in adult women. Paper presented at the annual meeting of the American Association for Health, Physical Education, and Recreation, Houston, Texas.

Henschel, A. 1971. The environment and performance. In Simonson, E. (ed.). *Physiology of Work Capacity and Fatigue*. Springfield, Ill.: Thomas, pp. 325–347.

Hermansen, L. 1969. Anaerobic energy release. *Medicine and Science in Sports* 1, 32–38.

———, and Wachtlová, M. 1971. Capillary density of skeletal muscle in well-trained and untrained men. *Journal of Applied Physiology* 30, 860–863.

Hettinger, T., and Müller, E. A. 1953. Muscle strength and training. *Arbeitsphysiologie (Work Physiology)* 15, 111–126.

———. 1961. *Physiology of Strength*. Springfield, Ill.: Thomas.

Hill, A. V. 1964. The effect of load on the heat of shortening muscle. *Proceedings of the Royal Society (Biology)* 159, 297–318.

Hill, S. R.; Goetz, F. C.; Fox, H. M. and others. 1956. Studies on adrenocortical and psychological responses to stress in man. *Archives of Internal Medicine* 97, 269–298.

Hobson, J. A. 1968. Sleep after exercise. *Science* 162, 1503–1505.

Holloszy, J. O. 1967. Effects of exercise on mitochondrial oxygen uptake and respiratory enzyme activity in skeletal muscle. *Journal of Biological Chemistry* 242, 2278–2282.

———; Oscai, L. B.; Molé, P. A.; and Don, I. J. 1971. Biochemical adaptations to endurance exercise in skeletal muscle. In Pernow, B., and Saltin, B. (eds.). *Muscle Metabolism During Exercise*. New York, Plenum, pp. 51–61.

Holmgren, A. 1967. Cardiorespiratory determinants of cardiovascular fitness. *Canadian Medical Association Journal* 96, 697–702.

———, and Astrand, P. O. 1966. D_L and the dimensions and functional capacities of the O_2 transport system in humans. *Journal of Applied Physiology* 21, 1463–1470.

Horstman, D., and Gleser, M. 1973. Maximum oxygen consumption and blood flow with reduced hemoglobin for dog skeletal muscle in situ. Paper read at the annual meeting of the American College of Sports Medicine, Seattle, Washington.

Hultman, E. 1971. Muscle glycogen stores and prolonged exercise. In Shephard, R. J. (ed.). *Frontiers of Fitness*. Springfield, Ill.: Thomas, pp. 37–60.

Hunter, W. M.; Fonseka, C. C.; and Passmore, R. 1965. Growth hormone: Important role in muscular exercise in adults. *Science* 150, 1051–1053.

Huxley, H. E. 1965. The mechanism of muscular contraction. *Scientific American* 213(6), 18–27.

Iatridis, S., and Ferguson, J. 1963. Effect of physical exercise on blood clotting and fibrinolysis. *Journal of Applied Physiology* 18, 337–344.

Ikai, M., and Steinhaus, A. H. 1961. Some factors modifying the expression of human strength. *Journal of Applied Physiology* 16, 157–163.

Ishmail, A. H., and Trachtman, J. 1973. Jogging the imagination. *Psychology Today* March, 79–82.

Issekutz, B., and Miller, H. 1962. Plasma free fatty acids during exercise and the effect of lactic acid. *Proceedings of the Society of Experimental Biology and Medicine* 110, 237–239.

———; Miller, H. I.; Paul, P.; and Rodahl, K. 1965. Aerobic work capacity and plasma FFA turnover. *Journal of Applied Physiology* 20, 293–296.

Jackson, D. 1971. The cloud comes to Quibbletown. *Life* 71 (24), 72–82.

Jackson, J.; Sharkey, B. J.; and Johnston, L. P. 1968. Cardiorespiratory adaptations to training at specified frequencies. *Research Quarterly* 39, 295–300.

Johnson, L. C., and O'Shea, J. P. 1969. Anabolic steroid: Effects on strength development. *Science* 164, 957–958.

———; Fisher, G.; Silvester, L. J.; and Hofheins, C. C. 1972. Anabolic steroid: Effects on strength, body weight, oxygen uptake, and spermatogenesis upon mature males. *Medicine and Science in Sports* 4, 43–45.

Johnson, R. L. 1967. Pulmonary diffusion as a limiting factor in exercise stress. In Chapman, C. B. (ed.). *Physiology of Muscular Exercise*. New York: American Heart Association, pp. 154–160.

Jokl, E. 1964. *Physiology of Exercise*. Springfield, Ill.: Thomas.

———, and McClellan, J. T. 1970. Exercise and cardiac death. *Journal of the American Medical Association* 213, 1489–1491.

Kane, M. 1971. An assessment of 'Black is Best'. *Sports Illustrated* 34, 72–83.

Karlsson, J., and Saltin, B. 1970. Lactate, ATP, and CP in working muscles during exhaustive exercise in man. *Journal of Applied Physiology* 29, 598–602.

———, and Saltin, B. 1971. Diet, muscle glycogen and endurance performance. *Journal of Applied Physiology* 31, 203–206.

Karpovich, P. V. 1934. Effects of oxygen inhalation on swimming performance. *Research Quarterly* 5, 24–30.

Karvonen, M. J.; Kentala, E.; and Mustala, O. 1957. The effects of training on heart rate: A longitudinal study. *Finland Journal of Experimental Medicine* 35, 307–315.

Katch, F. I., and Michael, E. D. 1971. Body composition of high school wrestlers according to age and wrestling weight category. *Medicine and Science in Sports* 3, 190–194.

Kattus, A. A.; Brock, L. L.; Bruce, R. A.; and others. 1972. Exercise testing and training of apparently healthy individuals: A handbook for physicians. New York: American Heart Association.

Keul, J. 1971. Myocardial metabolism in athletes. In Pernow, B. and Saltin, B. (eds.). *Muscle Metabolism During Exercise*. New York: Plenum, pp. 447–456.

Kiester, E. 1971. Their hearts are crying for help. *Family Health* 3, 16–19.

Klissouras, V. 1971. Heritability of adaptive variation. *Journal of Applied Physiology* 31, 338–344.

Lamb, D. R. 1968. Influence of exercise on bone growth and metabolism. *Kinesiology Review* 1, 43–48.

Lind, A. R.; McNicol, G. W.; and Donald, K. W. 1966. Circulatory adjustments to sustained (static) muscular activity. In Evang, K., and Andersen, K. L. *Physical Activity in Health and Disease*. Baltimore: Williams & Wilkins, pp. 38–63.

Londree, B. 1973. Personal communication.

McManus, B. M., and Lamb, D. R. 1972. Skeletal muscle amino acid incorporated and testosterone uptake in exercised guinea pigs. Paper presented at the annual meeting of the American College of Sports Medicine, Philadelphia.

Maksud, M. G.; Hamilton, L. H.; Coutts, K. D.; and Wiley, R. L. 1971. Pulmonary function measurements of Olympic speed skaters from the U.S. *Medicine and Science in Sports* 3, 66–71.

Malina, R. 1972. Anthropology, growth and physical education. In *Physical Education: An Interdisciplinary Approach*. New York: Macmillan, pp. 237–309.

Margaria, R.; Aghemo, P.; and Rovelli, E. 1966. Measurement of muscular power (anaerobic) in man. *Journal of Applied Physiology* 21, 1662–1664.

———, and Cerretelli, P. 1968. The respiratory system and exercise. In Falls, H. B. (ed.). *Exercise Physiology*. New York: Academic, pp. 43–78.

———; Di Prampero, P. E.; Aghemo, P.; Derevenco, P.; and Mariana, M. 1971. Effect of a steady-state exercise on maximal anaerobic power in man. *Journal of Applied Physiology* 30, 885–889.

Massey, B. H.; Nelson, R. C.; Sharkey, B. J.; and Comden, T. 1965. Effects of high-frequency electrical stimulation on the size and strength of skeletal muscle. *Journal of Sports Medicine* 5, 136–144.

Merton, P. A. 1954. Voluntary strength and fatigue. *Journal of Physiology* (London) 123, 553–564.

Miller, N. 1969. Learning of visceral and glandular responses. *Science* 163, 434–435.

Miller, R. E., and Mason, J. W. 1964. Changes in 17-hydroxy-corticosteroid excretion related to increased muscular work. In *Medical Aspects of Stress in the Military Climate*. Washington, D. C.: Walter Reed Army Inst. of Research, pp. 137–151.

Missiuro, W.; Kirschner, H.; and Kozlowski, S. 1962. Electromyographic manifestations of fatigue during work of different intensity. *Acta Physiologia Polonica* 13, 11–20.

Molé, P. A.; Baldwin, K. M.; Terjung, R. L.; and Holloszy, J. O. 1973. Enzymatic pathways of pyruvate metabolism in skeletal muscle: adaptations to exercise. *American Journal of Physiology* 224, 50–54.

———; Oscai, L. B.; and Holloszy, J. O. 1971. Adaptation of muscle to exercise: Increase in levels of palmityl CoA synthetase, carnitine palmityltransferase, and palmityl CoA dehydrogenase, and in the capacity to oxidize fatty acids. *Journal of Clinical Investigation* 50, 2323–2329.

Morris, J., and Crawford, M. 1958. Coronary heart disease and physical activity of work. *Journal of the British Medical Association* 2, 1485–1496.

Moxley, R. T.; Brakman, P.; and Astrup, T. 1970. Resting levels of fibrinolysis in blood in inactive and exercising men. *Journal of Applied Physiology* 28, 549–552.

Müller, E. A., and Rohmert, W. 1963. Die geschwindigkeit der muskelkraft zunahme bei isometrischen training (Augmentation of muscle strength with isometric training): *International Zeitschift Angewadte Physiologie* 19, 403–419.

Murray, J. M., and Weber, A. 1974. The cooperative action of muscle proteins. *Scientific American* 230, 58–71.

National Collegiate Athletic Association. 1972. Drugs and athletics do not mix. A pamphlet published in cooperation with the Bureau of Narcotics and Dangerous Drugs.

Oscai, L. B., and Holloszy, J. O. 1969. Effects of weight changes produced by exercise, food restriction, or overeating on body composition. *Journal of Clinical Investigation* 48, 2124–2128.

———; Molé, P. A.; and Holloszy, J. O. 1971. Effects of exercise on cardiac weight and mitochondria in male and female rats. *American Journal of Physiology* 220, 1944–1948.

Osnes, J., and Hermansen, L. 1972. Acid-base balance after maximal exercise of short duration. *Journal of Applied Physiology* 32, 59–63.

Palladin, A. V. 1945. The biochemistry of muscle training. *Science* 102, 576–578.

Pařízková, J.; Eiselt, E.; Šprynarová, Š.; and Wachtlová, M. 1971. Body composition, aerobic capacity, and density of muscle capillaries in young and old men. *Journal of Applied Physiology* 31, 323–325.

Passmore, R., and Durnin, J. V. G. A. 1955. Human energy expenditure. *Physiology Review* 35, 801–840.

Paul, P., and Issekutz, B. 1967. Role of extramuscular energy sources in the metabolism of the exercising dog. *Journal of Applied Physiology* 22, 615–622.

———. 1970. FFA metabolism of normal dogs during steady-state exercise at different work loads. *Journal of Applied Physiology* 28, 127–132.

Pernow, B., and Saltin, B. 1971. Availability of substrates and capacity for prolonged heavy exercise in man. *Journal of Applied Physiology* 31, 416–422.

Pollock, M. L.; Cureton, T. K.; and Greninger, M. S. 1969. Effects of frequency of training on working capacity, cardiovascular function and body composition of adult men. *Medicine and Science in Sports* 1, 70–74.

———; Dimmick, J.; Miller, H. S.; Kendrick, Z.; and Linnerud, A. C. 1972. Effects of mode of training on cardiovascular function and body composition of middle-aged men. Paper read at the annual meeting of the American College of Sports Medicine, Philadelphia.

———; Miller, H. S.; and Wilmore, J. 1972. Physiological characteristics of champion American track athletes 40 to 70 years of age. Paper read at the Scientific Congress in conjunction with the XX Olympiad, Munich, West Germany.

Poupa, O., and Rakušan, K. 1966. The terminal microcirculatory bed in the heart of athletic and non-athletic animals. In Evang, K., and Andersen, K. L. (eds.). *Physical Activity in Health and Disease*. Baltimore: Williams & Wilkins, pp. 18–29.

Raab, W. 1965. Prevention of ischaemic heart disease. *Medical Services Journal of Canada* 21, 719–734.

Roberts, T. W.; Smith, J. L.; and Roberts, E. M. 1970. Fusimotor neurons in voluntary movement. In Vendebregt, J.; Wartenweiler, J.; and Karger, S. (eds.). *Medicine and Sport: Biomechanics II*. Baltimore: University Park, pp. 48–54.

Robinson, S. 1967. Training, acclimatization and heat tolerance. *Canadian Medical Association Journal* 96, 795–799.

Rochelle, R. 1961. Blood plasma cholesterol changes during a physical training program. *Research Quarterly* 32, 538–550.

Rode, A., and Shephard, R. J. 1971. The influence of cigarette smoking upon the oxygen cost of breathing in near-maximal exercise. *Medicine and Science in Sports* 3, 51–55.

Romanul, F. C. A. 1971. Reversal of enzymatic profiles and capillary supply of muscle fibers in fast and slow muscles after cross innervation. In Pernow, B., and Saltin, B. (eds.). *Muscle Metabolism During Exercise*. New York: Plenum, 21–32.

Rosentsweieg, J., and Hinson, M. 1972. Comparative muscle action potential values of isometric, isotonic and isokinetic contraction. Paper read at the annual meeting of the American Association for Health, Physical Education and Recreation, Houston.

Roth, E. M. (ed.). 1968. *Compendium of Human Responses to the Aerospace Environment (III)*. Washington, D.C.: N.A.S.A.

Rowell, L. B. 1971. Cardiovascular limitations to work capacity. In Simonson, E. (ed.). *Physiology of Work Capacity and Fatigue*. Springfield, Ill.: Thomas, 132–169.

———; Taylor, H. L.; Wang, Y.; and Carlson, W. S. 1964. Saturation of arterial blood with oxygen during maximal exercise. *Journal of Applied Physiology* 19, 284–286.

Rushmer, R. F. 1961. *Cardiovascular Dynamics*. Philadelphia: Saunders.

Saltin, B. 1973. Metabolic fundamentals in exercise. *Medicine and Science in Sports* 5, 137–146.

———, and Astrand, P. O. 1967. Maximal oxygen uptake in athletes. *Journal of Applied Physiology* 23, 353–358.

Sandow, A.; Taylor, S. R.; and Preisier, H. 1965. Role of the action potential in excitation-contraction coupling. *Federation Proceedings* 24, 1116–1123.

Selye, H. *The Stress of Life*. 1956. New York: McGraw-Hill.

Sharkey, B. J. 1961. Oxygen inhalation and performance in swimming. *Swimming World* 3, 11.

———, and Holleman, J. P. 1967. Cardiorespiratory adaptations to training at specified intensities. *Research Quarterly* 38, 398–404.

———. 1970. Intensity and duration of training and the development of cardiorespiratory endurance. *Medicine and Science in Sports* 2, 197–202.

———. 1974. *Physiological Fitness and Weight Control*. Missoula, Mont.: Mountain Press.

Shaw, D., and MacNaughton, D. 1963. Relationship between blood fibrinolytic activity and body fatness. *Lancet* 1, 352–354.

Shephard, R. J. 1966. World standards of cardiorespiratory performance. *Archives of Environmental Health* 13, 664–672.

———. 1968. Intensity, duration and frequency of exercise as determinants of the response to a training regime. *International Zeitschift angewandte Physiologie* 26, 272–278.

Simonson, E. 1971. *Physiology of Work Capacity and Fatigue.* Springfield, Ill.: Thomas.

Skinner, J. S.; Holloszy, J.; Toro, G.; Barry, A.; and Cureton, T. K. 1967. Effects of a six-month program of endurance exercise on work tolerance, serum lipids and ULF-ballistocardiograms of fifteen middle-aged men. In Karvonen, M. J., and Barry, A. J. (eds.). *Physical Activity and the Heart.* Springfield, Ill.: Thomas, pp. 79–98.

Spain, D. M.; Nathan, D. J.; and Gellis, M. 1963. Weight, body type and prevalence of coronary atherosclerotic heart disease in males. *American Journal of Medical Science* 245, 63–72.

Steadman, R. T., and Sharkey, B. J. 1969. Exercise as a stressor. *Journal of Sports Medicine* 9, 230–235.

Stevenson, J.; Felek, V.; Rechnitzer, P.; and Beaton, J. 1964. Effect of exercise on coronary tree size in rats. *Circulation Research* 15, 265–270.

Stiles, M. H. 1969. A first-hand report of the Mexico City Olympic Games. *Newsletter of the American College of Sports Medicine* 4, 4–5.

Suzuki, T. 1967. Effects of muscular exercise on adrenal 17-hydroxycorticosteroid secretion in the dog. *Endocrinology* 80, 1148–1151.

Tanner, J. M. 1964. *The Physique of the Olympic Athlete.* London: Allen and Unwin.

Taylor, A. W.; Lappage, R.; and Rao, S. 1971. Skeletal muscle glycogen stores after submaximal and maximal work. *Medicine and Science in Sports* 3, 75–78.

——— (ed.). 1972. *Training: Scientific Basis and Application.* Springfield, Ill.: Thomas.

Tcheng, T., and Tipton, C. M. 1972. Tcheng-Tipton equations for predicting an optimum body weight for high school age wrestlers. Paper presented at the annual meeting of the American Association for Health, Physical Education and Recreation, Houston.

Thomas, C. L. 1969. Effect of vigorous athletic activity on women. In American Academy of Orthopedic Surgeons. *Symposium on Sports Medicine.* St. Louis: Mosby, pp. 128–135.

Thompson, R. F. 1967. *Foundations of Physiological Psychology.* New York: Harper & Row.

Thompson, S. H., and Sharkey, B. J. 1966. Physiological cost and air flow resistance of respiratory protective devices. *Ergonomics* 9, 495–499.

Thys, H.; Faraggiana, T.; and Margaria, R. 1972. Utilization of muscle elasticity in exercise. *Journal of Applied Physiology* 32, 491–494.

U. S. Navy Diving Manual. 1970. Washington, D.C.: Department of the Navy.

Van Linge, B. 1962. The response of muscle to strenuous exercise. *Journal of Bone and Joint Surgery* 44, 711–721.

Wegmann, H. M. 1966. Enzymatic and hormonal responses to exercise, lowered pressure, and acceleration. *Federation Proceedings* 25, 1405–1408.

Weiser, P. C. 1971. Alterations in enzyme activity in work and fatigue. In Simonson, E. (ed.). *Physiology of Work Capacity and Fatigue*. Springfield, Ill.: Thomas, pp. 97–131.

Whiddon, T. R.; Sharkey, B. J.; and Steadman, R. J. 1969. Exercise, stress and blood clotting in men. *Research Quarterly* 40, 431–434.

Williams, R. J. 1967. *You Are Extraordinary*. New York: Random House.

Wilmore, J., and Haskell, W. 1971. Use of the heart rate-energy expenditure relationship in the individualized prescription of exercise. *American Journal of Clinical Nutrition* 24, 1186–1192.

Wilmore, J. H., and Behnke, A. R. 1970. An anthropometric estimation of body density and lean body weight in young women. *American Journal of Clinical Nutrition* 23, 267–274.

Wilt, F. 1968. Training for competitive running. In Falls, H. B. (ed.). *Exercise Physiology*. New York: Academic, pp. 395–414.

Yoshimura, H. 1965. Studies on protein metabolism in hard muscular work in relation to its nutritional requirement. In Vaughan, L. (ed.). *Nutritional Requirements for Survival in the Cold and at Altitude*. Fort Wainwright, Alaska: Arctic Aeromedical Laboratory, pp. 85–120.

Zauner, C. W.; Sterling, L. F.; Dunavant, B. G.; and Roessler, G. S. 1971. Relationship among fitness parameters, body composition, age, habitual activity and postprandial lipemia. *Medicine and Science in Sports* 3, 118–121.

Zukel, W.; Lewis, R. H.; and Enterline, P. 1959. A short-term community study of the epidemiology of coronary heart disease. *American Journal of Public Health* 49, 1630–1639.

INDEX

74 75 76 7 6 5 4 3 2 1